The Org Manual

Release 9.1.6 (release_9.1.6)

by Carsten Dominik

with contributions by Bastien Guerry, Nicolas Goaziou, Eric Schulte, Jambunathan K, Dan Davison, Thomas Dye, David O'Toole, and Philip Rooke.

This manual is for Org version 9.1.6 (release_9.1.6).

Short Contents

Table of Contents

1 Introduction

1.1 Summary

Org is a mode for keeping notes, maintaining TODO lists, and project planning with a fast and effective plain-text system. It also is an authoring system with unique support for literate programming and reproducible research.

Org is implemented on top of Outline mode, which makes it possible to keep the content of large files well structured. Visibility cycling and structure editing help to work with the tree. Tables are easily created with a built-in table editor. Plain text URL-like links connect to websites, emails, Usenet messages, BBDB entries, and any files related to the projects.

Org develops organizational tasks around notes files that contain lists or information about projects as plain text. Project planning and task management makes use of metadata which is part of an outline node. Based on this data, specific entries can be extracted in queries and create dynamic *agenda views* that also integrate the Emacs calendar and diary. Org can be used to implement many different project planning schemes, such as David Allen's GTD system.

Org files can serve as a single source authoring system with export to many different formats such as HTML, LaTeX, Open Document, and Markdown. New export backends can be derived from existing ones, or defined from scratch.

Org files can include source code blocks, which makes Org uniquely suited for authoring technical documents with code examples. Org source code blocks are fully functional; they can be evaluated in place and their results can be captured in the file. This makes it possible to create a single file reproducible research compendium.

Org keeps simple things simple. When first fired up, it should feel like a straightforward, easy to use outliner. Complexity is not imposed, but a large amount of functionality is available when needed. Org is a toolbox. Many users actually run only a (very personal) fraction of Org's capabilities, and know that there is more whenever they need it.

All of this is achieved with strictly plain text files, the most portable and future-proof file format. Org runs in Emacs. Emacs is one of the most widely ported programs, so that Org mode is available on every major platform.

There is a website for Org which provides links to the newest version of Org, as well as additional information, frequently asked questions (FAQ), links to tutorials, etc. This page is located at `http://orgmode.org`.

An earlier version (7.3) of this manual is available as a paperback book from Network Theory Ltd.

1.2 Installation

Org is part of recent distributions of GNU Emacs, so you normally don't need to install it. If, for one reason or another, you want to install Org on top of this pre-packaged version, there are three ways to do it:

- By using Emacs package system.
- By downloading Org as an archive.
- By using Org's git repository.

We **strongly recommend** to stick to a single installation method.

Using Emacs packaging system

Recent Emacs distributions include a packaging system which lets you install Elisp libraries. You can install Org with *M-x package-install RET org*.

Important: you need to do this in a session where no `.org` file has been visited, i.e., where no Org built-in function have been loaded. Otherwise autoload Org functions will mess up the installation.

Then, to make sure your Org configuration is taken into account, initialize the package system with `(package-initialize)` in your Emacs init file before setting any Org option. If you want to use Org's package repository, check out the Org ELPA page.

Downloading Org as an archive

You can download Org latest release from Org's website. In this case, make sure you set the load-path correctly in your Emacs init file:

```
(add-to-list 'load-path "~/path/to/orgdir/lisp")
```

The downloaded archive contains contributed libraries that are not included in Emacs. If you want to use them, add the `contrib` directory to your load-path:

```
(add-to-list 'load-path "~/path/to/orgdir/contrib/lisp" t)
```

Optionally, you can compile the files and/or install them in your system. Run `make help` to list compilation and installation options.

Using Org's git repository

You can clone Org's repository and install Org like this:

```
$ cd ~/src/
$ git clone git://orgmode.org/org-mode.git
$ make autoloads
```

Note that in this case, `make autoloads` is mandatory: it defines Org's version in `org-version.el` and Org's autoloads in `org-loaddefs.el`.

Remember to add the correct load-path as described in the method above.

You can also compile with `make`, generate the documentation with `make doc`, create a local configuration with `make config` and install Org with `make install`. Please run `make help` to get the list of compilation/installation options.

For more detailed explanations on Org's build system, please check the Org Build System page on Worg.

1.3 Activation

Org mode buffers need font-lock to be turned on: this is the default in Emacs[1].

There are compatibility issues between Org mode and some other Elisp packages, please take the time to check the list (see Section 15.10.2 [Conflicts], page 238).

The four Org commands `org-store-link`, `org-capture`, `org-agenda`, and `org-iswitchb` should be accessible through global keys (i.e., anywhere in Emacs, not just in Org buffers). Here are suggested bindings for these keys, please modify the keys to your own liking.

```
(global-set-key "\C-cl" 'org-store-link)
(global-set-key "\C-ca" 'org-agenda)
(global-set-key "\C-cc" 'org-capture)
(global-set-key "\C-cb" 'org-iswitchb)
```

Files with the `.org` extension use Org mode by default. To turn on Org mode in a file that does not have the extension `.org`, make the first line of a file look like this:

```
MY PROJECTS    -*- mode: org; -*-
```

which will select Org mode for this buffer no matter what the file's name is. See also the variable `org-insert-mode-line-in-empty-file`.

Many commands in Org work on the region if the region is *active*. To make use of this, you need to have `transient-mark-mode` turned on, which is the default. If you do not like `transient-mark-mode`, you can create an active region by using the mouse to select a region, or pressing *C-SPC* twice before moving the cursor.

1.4 Feedback

If you find problems with Org, or if you have questions, remarks, or ideas about it, please mail to the Org mailing list `emacs-orgmode@gnu.org`. You can subscribe to the list on this web page. If you are not a member of the mailing list, your mail will be passed to the list after a moderator has approved it[2].

For bug reports, please first try to reproduce the bug with the latest version of Org available—if you are running an outdated version, it is quite possible that the bug has been fixed already. If the bug persists, prepare a report and provide as much information as possible, including the version information of Emacs (*M-x emacs-version RET*) and Org (*M-x org-version RET*), as well as the Org related setup in the Emacs init file. The easiest way to do this is to use the command

M-x org-submit-bug-report RET

which will put all this information into an Emacs mail buffer so that you only need to add your description. If you are not sending the Email from within Emacs, please copy and paste the content into your Email program.

Sometimes you might face a problem due to an error in your Emacs or Org mode setup. Before reporting a bug, it is very helpful to start Emacs with minimal customizations and

---

[1] If you don't use font-lock globally, turn it on in Org buffer with `(add-hook 'org-mode-hook 'turn-on-font-lock)`

[2] Please consider subscribing to the mailing list, in order to minimize the work the mailing list moderators have to do.

reproduce the problem. Doing so often helps you determine if the problem is with your customization or with Org mode itself. You can start a typical minimal session with a command like the example below.

```
$ emacs -Q -l /path/to/minimal-org.el
```

However if you are using Org mode as distributed with Emacs, a minimal setup is not necessary. In that case it is sufficient to start Emacs as `emacs -Q`. The `minimal-org.el` setup file can have contents as shown below.

```
;;; Minimal setup to load latest 'org-mode'

;; activate debugging
(setq debug-on-error t
      debug-on-signal nil
      debug-on-quit nil)

;; add latest org-mode to load path
(add-to-list 'load-path "/path/to/org-mode/lisp")
(add-to-list 'load-path "/path/to/org-mode/contrib/lisp" t)
```

If an error occurs, a backtrace can be very useful (see below on how to create one). Often a small example file helps, along with clear information about:

1. What exactly did you do?
2. What did you expect to happen?
3. What happened instead?

Thank you for helping to improve this program.

How to create a useful backtrace

If working with Org produces an error with a message you don't understand, you may have hit a bug. The best way to report this is by providing, in addition to what was mentioned above, a *backtrace*. This is information from the built-in debugger about where and how the error occurred. Here is how to produce a useful backtrace:

1. Reload uncompiled versions of all Org mode Lisp files. The backtrace contains much more information if it is produced with uncompiled code. To do this, use

   C-u M-x org-reload RET

   or select `Org -> Refresh/Reload -> Reload Org uncompiled` from the menu.
2. Go to the `Options` menu and select `Enter Debugger on Error`.
3. Do whatever you have to do to hit the error. Don't forget to document the steps you take.
4. When you hit the error, a `*Backtrace*` buffer will appear on the screen. Save this buffer to a file (for example using *C-x C-w*) and attach it to your bug report.

1.5 Typesetting conventions used in this manual

TODO keywords, tags, properties, etc.

Org mainly uses three types of keywords: TODO keywords, tags and property names. In this manual we use the following conventions:

`TODO`
`WAITING` TODO keywords are written with all capitals, even if they are user-defined.

`boss`
`ARCHIVE` User-defined tags are written in lowercase; built-in tags with special meaning are written with all capitals.

`Release`
`PRIORITY` User-defined properties are capitalized; built-in properties with special meaning are written with all capitals.

Moreover, Org uses *option keywords* (like `#+TITLE` to set the title) and *environment keywords* (like `#+BEGIN_EXPORT html` to start a `HTML` environment). They are written in uppercase in the manual to enhance its readability, but you can use lowercase in your Org file.

Key bindings and commands

The manual suggests a few global key bindings, in particular *C-c a* for `org-agenda` and *C-c c* for `org-capture`. These are only suggestions, but the rest of the manual assumes that these key bindings are in place in order to list commands by key access.

Also, the manual lists both the keys and the corresponding commands for accessing a functionality. Org mode often uses the same key for different functions, depending on context. The command that is bound to such keys has a generic name, like `org-metaright`. In the manual we will, wherever possible, give the function that is internally called by the generic command. For example, in the chapter on document structure, *M-right* will be listed to call `org-do-demote`, while in the chapter on tables, it will be listed to call `org-table-move-column-right`. If you prefer, you can compile the manual without the command names by unsetting the flag `cmdnames` in `org.texi`.

2 Document structure

Org is based on Outline mode and provides flexible commands to edit the structure of the document.

2.1 Outlines

Org is implemented on top of Outline mode. Outlines allow a document to be organized in a hierarchical structure, which (at least for me) is the best representation of notes and thoughts. An overview of this structure is achieved by folding (hiding) large parts of the document to show only the general document structure and the parts currently being worked on. Org greatly simplifies the use of outlines by compressing the entire show/hide functionality into a single command, `org-cycle`, which is bound to the `TAB` key.

2.2 Headlines

Headlines define the structure of an outline tree. The headlines in Org start with one or more stars, on the left margin[1][2]. For example:

```
* Top level headline
** Second level
*** 3rd level
    some text
*** 3rd level
    more text

* Another top level headline
```

Note that a headline named after `org-footnote-section`, which defaults to '`Footnotes`', is considered as special. A subtree with this headline will be silently ignored by exporting functions.

Some people find the many stars too noisy and would prefer an outline that has whitespace followed by a single star as headline starters. Section 15.8 [Clean view], page 234, describes a setup to realize this.

An empty line after the end of a subtree is considered part of it and will be hidden when the subtree is folded. However, if you leave at least two empty lines, one empty line will remain visible after folding the subtree, in order to structure the collapsed view. See the variable `org-cycle-separator-lines` to modify this behavior.

2.3 Visibility cycling

2.3.1 Global and local cycling

Outlines make it possible to hide parts of the text in the buffer. Org uses just two commands, bound to `TAB` and *`S-TAB`* to change the visibility in the buffer.

---

[1] See the variables `org-special-ctrl-a/e`, `org-special-ctrl-k`, and `org-ctrl-k-protect-subtree` to configure special behavior of *C-a*, *C-e*, and *C-k* in headlines.

[2] Clocking only works with headings indented less than 30 stars.

`TAB` `org-cycle`
Subtree cycling: Rotate current subtree among the states

```
,-> FOLDED -> CHILDREN -> SUBTREE --.
'-----------------------------------'
```

The cursor must be on a headline for this to work[3].

S-TAB `org-global-cycle`
C-u `TAB` *Global cycling*: Rotate the entire buffer among the states

```
,-> OVERVIEW -> CONTENTS -> SHOW ALL --.
'--------------------------------------'
```

When *S-TAB* is called with a numeric prefix argument N, the CONTENTS view up to headlines of level N will be shown. Note that inside tables, *S-TAB* jumps to the previous field.

You can run global cycling using `TAB` only if point is at the very beginning of the buffer, but not on a headline, and `org-cycle-global-at-bob` is set to a non-`nil` value.

C-u C-u TAB `org-set-startup-visibility`
Switch back to the startup visibility of the buffer (see Section 2.3.2 [Initial visibility], page 8).

C-u C-u C-u TAB `outline-show-all`
Show all, including drawers.

C-c C-r `org-reveal`
Reveal context around point, showing the current entry, the following heading and the hierarchy above. Useful for working near a location that has been exposed by a sparse tree command (see Section 2.6 [Sparse trees], page 11) or an agenda command (see Section 10.5 [Agenda commands], page 115). With a prefix argument show, on each level, all sibling headings. With a double prefix argument, also show the entire subtree of the parent.

C-c C-k `outline-show-branches`
Expose all the headings of the subtree, CONTENTS view for just one subtree.

C-c TAB `outline-show-children`
Expose all direct children of the subtree. With a numeric prefix argument N, expose all children down to level N.

C-c C-x b `org-tree-to-indirect-buffer`
Show the current subtree in an indirect buffer[4]. With a numeric prefix argument N, go up to level N and then take that tree. If N is negative then go up that many levels. With a *C-u* prefix, do not remove the previously used indirect buffer.

---

[3] see, however, the option `org-cycle-emulate-tab`.

[4] The indirect buffer (see Section "Indirect Buffers" in *GNU Emacs Manual*) will contain the entire buffer, but will be narrowed to the current tree. Editing the indirect buffer will also change the original buffer, but without affecting visibility in that buffer.

C-c C-x v `org-copy-visible`

Copy the *visible* text in the region into the kill ring.

2.3.2 Initial visibility

When Emacs first visits an Org file, the global state is set to OVERVIEW, i.e., only the top level headlines are visible[5]. This can be configured through the variable `org-startup-folded`, or on a per-file basis by adding one of the following lines anywhere in the buffer:

```
#+STARTUP: overview
#+STARTUP: content
#+STARTUP: showall
#+STARTUP: showeverything
```

Furthermore, any entries with a '`VISIBILITY`' property (see Chapter 7 [Properties and columns], page 64) will get their visibility adapted accordingly. Allowed values for this property are `folded`, `children`, `content`, and `all`.

C-u C-u TAB `org-set-startup-visibility`

Switch back to the startup visibility of the buffer, i.e., whatever is requested by startup options and '`VISIBILITY`' properties in individual entries.

2.3.3 Catching invisible edits

Sometimes you may inadvertently edit an invisible part of the buffer and be confused on what has been edited and how to undo the mistake. Setting `org-catch-invisible-edits` to non-`nil` will help prevent this. See the docstring of this option on how Org should catch invisible edits and process them.

2.4 Motion

The following commands jump to other headlines in the buffer.

C-c C-n `org-next-visible-heading`

Next heading.

C-c C-p `org-previous-visible-heading`

Previous heading.

C-c C-f `org-forward-same-level`

Next heading same level.

C-c C-b `org-backward-same-level`

Previous heading same level.

C-c C-u `outline-up-heading`

Backward to higher level heading.

C-c C-j `org-goto`

Jump to a different place without changing the current outline visibility. Shows the document structure in a temporary buffer, where you can use the following keys to find your destination:

---

[5] When `org-agenda-inhibit-startup` is non-`nil`, Org will not honor the default visibility state when first opening a file for the agenda (see Section A.9 [Speeding up your agendas], page 249).

`TAB`	Cycle visibility.
`down / up`	Next/previous visible headline.
`RET`	Select this location.
`/`	Do a Sparse-tree search

The following keys work if you turn off `org-goto-auto-isearch`

`n / p`	Next/previous visible headline.
`f / b`	Next/previous headline same level.
`u`	One level up.
`0-9`	Digit argument.
`q`	Quit

See also the option `org-goto-interface`.

2.5 Structure editing

M-RET `org-meta-return`

Insert a new heading, item or row.

If the command is used at the *beginning* of a line, and if there is a heading or a plain list item (see Section 2.7 [Plain lists], page 12) at point, the new heading/item is created *before* the current line. When used at the beginning of a regular line of text, turn that line into a heading.

When this command is used in the middle of a line, the line is split and the rest of the line becomes the new item or headline. If you do not want the line to be split, customize `org-M-RET-may-split-line`.

Calling the command with a *C-u* prefix unconditionally inserts a new heading at the end of the current subtree, thus preserving its contents. With a double *C-u C-u* prefix, the new heading is created at the end of the parent subtree instead.

C-RET `org-insert-heading-respect-content`

Insert a new heading at the end of the current subtree.

M-S-RET `org-insert-todo-heading`

Insert new TODO entry with same level as current heading. See also the variable `org-treat-insert-todo-heading-as-state-change`.

C-S-RET `org-insert-todo-heading-respect-content`

Insert new TODO entry with same level as current heading. Like *C-RET*, the new headline will be inserted after the current subtree.

`TAB` `org-cycle`

In a new entry with no text yet, the first `TAB` demotes the entry to become a child of the previous one. The next `TAB` makes it a parent, and so on, all the way to top level. Yet another `TAB`, and you are back to the initial level.

M-left `org-do-promote`

Promote current heading by one level.

M-right `org-do-demote`

Demote current heading by one level.

M-S-left `org-promote-subtree`
: Promote the current subtree by one level.

M-S-right `org-demote-subtree`
: Demote the current subtree by one level.

M-up `org-move-subtree-up`
: Move subtree up (swap with previous subtree of same level).

M-down `org-move-subtree-down`
: Move subtree down (swap with next subtree of same level).

M-h `org-mark-element`
: Mark the element at point. Hitting repeatedly will mark subsequent elements of the one just marked. E.g., hitting `M-h` on a paragraph will mark it, hitting `M-h` immediately again will mark the next one.

C-c @ `org-mark-subtree`
: Mark the subtree at point. Hitting repeatedly will mark subsequent subtrees of the same level than the marked subtree.

C-c C-x C-w `org-cut-subtree`
: Kill subtree, i.e., remove it from buffer but save in kill ring. With a numeric prefix argument N, kill N sequential subtrees.

C-c C-x M-w `org-copy-subtree`
: Copy subtree to kill ring. With a numeric prefix argument N, copy the N sequential subtrees.

C-c C-x C-y `org-paste-subtree`
: Yank subtree from kill ring. This does modify the level of the subtree to make sure the tree fits in nicely at the yank position. The yank level can also be specified with a numeric prefix argument, or by yanking after a headline marker like '`****`'.

C-y `org-yank`
: Depending on the options `org-yank-adjusted-subtrees` and `org-yank-folded-subtrees`, Org's internal `yank` command will paste subtrees folded and in a clever way, using the same command as *C-c C-x C-y*. With the default settings, no level adjustment will take place, but the yanked tree will be folded unless doing so would swallow text previously visible. Any prefix argument to this command will force a normal `yank` to be executed, with the prefix passed along. A good way to force a normal yank is *C-u C-y*. If you use `yank-pop` after a yank, it will yank previous kill items plainly, without adjustment and folding.

C-c C-x c `org-clone-subtree-with-time-shift`
: Clone a subtree by making a number of sibling copies of it. You will be prompted for the number of copies to make, and you can also specify if any timestamps in the entry should be shifted. This can be useful, for example, to create a number of tasks related to a series of lectures to prepare. For more details, see the docstring of the command `org-clone-subtree-with-time-shift`.

`C-c C-w` **org-refile**
: Refile entry or region to a different location. See Section 9.5 [Refile and copy], page 99.

`C-c ^` **org-sort**
: Sort same-level entries. When there is an active region, all entries in the region will be sorted. Otherwise the children of the current headline are sorted. The command prompts for the sorting method, which can be alphabetically, numerically, by time (first timestamp with active preferred, creation time, scheduled time, deadline time), by priority, by TODO keyword (in the sequence the keywords have been defined in the setup) or by the value of a property. Reverse sorting is possible as well. You can also supply your own function to extract the sorting key. With a *C-u* prefix, sorting will be case-sensitive.

`C-x n s` **org-narrow-to-subtree**
: Narrow buffer to current subtree.

`C-x n b` **org-narrow-to-block**
: Narrow buffer to current block.

`C-x n w` **widen**
: Widen buffer to remove narrowing.

`C-c *` **org-toggle-heading**
: Turn a normal line or plain list item into a headline (so that it becomes a subheading at its location). Also turn a headline into a normal line by removing the stars. If there is an active region, turn all lines in the region into headlines. If the first line in the region was an item, turn only the item lines into headlines. Finally, if the first line is a headline, remove the stars from all headlines in the region.

When there is an active region (Transient Mark mode), promotion and demotion work on all headlines in the region. To select a region of headlines, it is best to place both point and mark at the beginning of a line, mark at the beginning of the first headline, and point at the line just after the last headline to change. Note that when the cursor is inside a table (see Chapter 3 [Tables], page 19), the Meta-Cursor keys have different functionality.

2.6 Sparse trees

An important feature of Org mode is the ability to construct *sparse trees* for selected information in an outline tree, so that the entire document is folded as much as possible, but the selected information is made visible along with the headline structure above it[6]. Just try it out and you will see immediately how it works.

Org mode contains several commands for creating such trees, all these commands can be accessed through a dispatcher:

`C-c /` **org-sparse-tree**
: This prompts for an extra key to select a sparse-tree creating command.

---

[6] See also the variable `org-show-context-detail` to decide how much context is shown around each match.

`C-c / r` or `C-c / /` — `org-occur`

Prompts for a regexp and shows a sparse tree with all matches. If the match is in a headline, the headline is made visible. If the match is in the body of an entry, headline and body are made visible. In order to provide minimal context, also the full hierarchy of headlines above the match is shown, as well as the headline following the match. Each match is also highlighted; the highlights disappear when the buffer is changed by an editing command[7], or by pressing `C-c C-c`. When called with a `C-u` prefix argument, previous highlights are kept, so several calls to this command can be stacked.

`M-g n` or `M-g M-n` — `next-error`

Jump to the next sparse tree match in this buffer.

`M-g p` or `M-g M-p` — `previous-error`

Jump to the previous sparse tree match in this buffer.

For frequently used sparse trees of specific search strings, you can use the option `org-agenda-custom-commands` to define fast keyboard access to specific sparse trees. These commands will then be accessible through the agenda dispatcher (see Section 10.2 [Agenda dispatcher], page 103). For example:

```
(setq org-agenda-custom-commands
      '(("f" occur-tree "FIXME")))
```

will define the key `C-c a f` as a shortcut for creating a sparse tree matching the string '`FIXME`'.

The other sparse tree commands select headings based on TODO keywords, tags, or properties and will be discussed later in this manual.

To print a sparse tree, you can use the Emacs command `ps-print-buffer-with-faces` which does not print invisible parts of the document. Or you can use `C-c C-e C-v` to export only the visible part of the document and print the resulting file.

2.7 Plain lists

Within an entry of the outline tree, hand-formatted lists can provide additional structure. They also provide a way to create lists of checkboxes (see Section 5.6 [Checkboxes], page 56). Org supports editing such lists, and every exporter (see Chapter 12 [Exporting], page 137) can parse and format them.

Org knows ordered lists, unordered lists, and description lists.

- *Unordered* list items start with '`-`', '`+`', or '`*`'[8] as bullets.
- *Ordered* list items start with a numeral followed by either a period or a right parenthesis[9], such as '`1.`' or '`1)`'[10]. If you want a list to start with a different value (e.g., 20),

[7] This depends on the option `org-remove-highlights-with-change`

[8] When using '`*`' as a bullet, lines must be indented or they will be seen as top-level headlines. Also, when you are hiding leading stars to get a clean outline view, plain list items starting with a star may be hard to distinguish from true headlines. In short: even though '`*`' is supported, it may be better to not use it for plain list items.

[9] You can filter out any of them by configuring `org-plain-list-ordered-item-terminator`.

[10] You can also get '`a.`', '`A.`', '`a)`' and '`A)`' by configuring `org-list-allow-alphabetical`. To minimize confusion with normal text, those are limited to one character only. Beyond that limit, bullets will automatically fallback to numbers.

start the text of the item with `[@20]`[11]. Those constructs can be used in any item of the list in order to enforce a particular numbering.

- *Description* list items are unordered list items, and contain the separator ' `::` ' to distinguish the description *term* from the description.

Items belonging to the same list must have the same indentation on the first line. In particular, if an ordered list reaches number '`10.`', then the 2–digit numbers must be written left-aligned with the other numbers in the list. An item ends before the next line that is less or equally indented than its bullet/number.

A list ends whenever every item has ended, which means before any line less or equally indented than items at top level. It also ends before two blank lines. In that case, all items are closed. Here is an example:

```
** Lord of the Rings
   My favorite scenes are (in this order)
   1. The attack of the Rohirrim
   2. Eowyn's fight with the witch king
      + this was already my favorite scene in the book
      + I really like Miranda Otto.
   3. Peter Jackson being shot by Legolas
      - on DVD only
      He makes a really funny face when it happens.
   But in the end, no individual scenes matter but the film as a whole.
   Important actors in this film are:
   - Elijah Wood :: He plays Frodo
   - Sean Astin :: He plays Sam, Frodo's friend.  I still remember
     him very well from his role as Mikey Walsh in The Goonies.
```

Org supports these lists by tuning filling and wrapping commands to deal with them correctly, and by exporting them properly (see Chapter 12 [Exporting], page 137). Since indentation is what governs the structure of these lists, many structural constructs like `#+BEGIN_...` blocks can be indented to signal that they belong to a particular item.

If you find that using a different bullet for a sub-list (than that used for the current list-level) improves readability, customize the variable `org-list-demote-modify-bullet`. To get a greater difference of indentation between items and their sub-items, customize `org-list-indent-offset`.

The following commands act on items when the cursor is in the first line of an item (the line with the bullet or number). Some of them imply the application of automatic rules to keep list structure intact. If some of these actions get in your way, configure `org-list-automatic-rules` to disable them individually.

`TAB` `org-cycle`
: Items can be folded just like headline levels. Normally this works only if the cursor is on a plain list item. For more details, see the variable `org-cycle-include-plain-lists`. If this variable is set to `integrate`, plain list items will be treated like low-level headlines. The level of an item is then given by

---

[11] If there's a checkbox in the item, the cookie must be put *before* the checkbox. If you have activated alphabetical lists, you can also use counters like `[@b]`.

the indentation of the bullet/number. Items are always subordinate to real headlines, however; the hierarchies remain completely separated. In a new item with no text yet, the first `TAB` demotes the item to become a child of the previous one. Subsequent `TAB`s move the item to meaningful levels in the list and eventually get it back to its initial position.

M-RET `org-insert-heading`
Insert new item at current level. With a prefix argument, force a new heading (see Section 2.5 [Structure editing], page 9). If this command is used in the middle of an item, that item is *split* in two, and the second part becomes the new item[12]. If this command is executed *before item's body*, the new item is created *before* the current one.

M-S-RET Insert a new item with a checkbox (see Section 5.6 [Checkboxes], page 56).

S-up
S-down Jump to the previous/next item in the current list[13], but only if `org-support-shift-select` is off. If not, you can still use paragraph jumping commands like *C-up* and *C-down* to quite similar effect.

M-up
M-down Move the item including subitems up/down[14] (swap with previous/next item of same indentation). If the list is ordered, renumbering is automatic.

M-left
M-right Decrease/increase the indentation of an item, leaving children alone.

M-S-left
M-S-right
Decrease/increase the indentation of the item, including subitems. Initially, the item tree is selected based on current indentation. When these commands are executed several times in direct succession, the initially selected region is used, even if the new indentation would imply a different hierarchy. To use the new hierarchy, break the command chain with a cursor motion or so.

As a special case, using this command on the very first item of a list will move the whole list. This behavior can be disabled by configuring `org-list-automatic-rules`. The global indentation of a list has no influence on the text *after* the list.

C-c C-c If there is a checkbox (see Section 5.6 [Checkboxes], page 56) in the item line, toggle the state of the checkbox. In any case, verify bullets and indentation consistency in the whole list.

C-c - Cycle the entire list level through the different itemize/enumerate bullets ('`-`', '`+`', '`*`', '`1.`', '`1)`') or a subset of them, depending on `org-plain-list-ordered-item-terminator`, the type of list, and its indentation. With a numeric prefix argument N, select the Nth bullet from this list. If there is an active region

---

[12] If you do not want the item to be split, customize the variable `org-M-RET-may-split-line`.

[13] If you want to cycle around items that way, you may customize `org-list-use-circular-motion`.

[14] See `org-list-use-circular-motion` for a cyclic behavior.

when calling this, all selected lines are converted to list items. With a prefix argument, selected text is changed into a single item. If the first line already was a list item, any item marker will be removed from the list. Finally, even without an active region, a normal line will be converted into a list item.

*C-c **
: Turn a plain list item into a headline (so that it becomes a subheading at its location). See Section 2.5 [Structure editing], page 9, for a detailed explanation.

*C-c C-**
: Turn the whole plain list into a subtree of the current heading. Checkboxes (see Section 5.6 [Checkboxes], page 56) will become TODO (resp. DONE) keywords when unchecked (resp. checked).

S-left/right
: This command also cycles bullet styles when the cursor in on the bullet or anywhere in an item line, details depending on `org-support-shift-select`.

C-c ^
: Sort the plain list. You will be prompted for the sorting method: numerically, alphabetically, by time, by checked status for check lists, or by a custom function.

2.8 Drawers

Sometimes you want to keep information associated with an entry, but you normally don't want to see it. For this, Org mode has *drawers*. They can contain anything but a headline and another drawer. Drawers look like this:

```
** This is a headline
   Still outside the drawer
   :DRAWERNAME:
   This is inside the drawer.
   :END:
   After the drawer.
```

You can interactively insert drawers at point by calling `org-insert-drawer`, which is bound to `C-c C-x d`. With an active region, this command will put the region inside the drawer. With a prefix argument, this command calls `org-insert-property-drawer` and add a property drawer right below the current headline. Completion over drawer keywords is also possible using *M-TAB*[15].

Visibility cycling (see Section 2.3 [Visibility cycling], page 6) on the headline will hide and show the entry, but keep the drawer collapsed to a single line. In order to look inside the drawer, you need to move the cursor to the drawer line and press `TAB` there. Org mode uses the `PROPERTIES` drawer for storing properties (see Chapter 7 [Properties and columns], page 64), and you can also arrange for state change notes (see Section 5.3.2 [Tracking TODO state changes], page 52) and clock times (see Section 8.4 [Clocking work time], page 80) to be stored in a drawer `LOGBOOK`. If you want to store a quick note in the LOGBOOK drawer, in a similar way to state changes, use

C-c C-z
: Add a time-stamped note to the LOGBOOK drawer.

---

[15] Many desktops intercept *M-TAB* to switch windows. Use *C-M-i* or *ESC TAB* instead for completion (see Section 15.1 [Completion], page 228).

You can select the name of the drawers which should be exported with `org-export-with-drawers`. In that case, drawer contents will appear in export output. Property drawers are not affected by this variable: configure `org-export-with-properties` instead.

2.9 Blocks

Org mode uses begin...end blocks for various purposes from including source code examples (see Section 11.5 [Literal examples], page 131) to capturing time logging information (see Section 8.4 [Clocking work time], page 80). These blocks can be folded and unfolded by pressing TAB in the begin line. You can also get all blocks folded at startup by configuring the option `org-hide-block-startup` or on a per-file basis by using

```
#+STARTUP: hideblocks
#+STARTUP: nohideblocks
```

2.10 Footnotes

Org mode supports the creation of footnotes.

A footnote is started by a footnote marker in square brackets in column 0, no indentation allowed. It ends at the next footnote definition, headline, or after two consecutive empty lines. The footnote reference is simply the marker in square brackets, inside text. Markers always start with `fn:`. For example:

```
The Org homepage[fn:1] now looks a lot better than it used to.
...
[fn:1] The link is: http://orgmode.org
```

Org mode extends the number-based syntax to *named* footnotes and optional inline definition. Here are the valid references:

`[fn:name]`
: A named footnote reference, where `name` is a unique label word, or, for simplicity of automatic creation, a number.

`[fn::This is the inline definition of this footnote]`
: A LaTeX-like anonymous footnote where the definition is given directly at the reference point.

`[fn:name:a definition]`
: An inline definition of a footnote, which also specifies a name for the note. Since Org allows multiple references to the same note, you can then use `[fn:name]` to create additional references.

Footnote labels can be created automatically, or you can create names yourself. This is handled by the variable `org-footnote-auto-label` and its corresponding `#+STARTUP` keywords. See the docstring of that variable for details.

The following command handles footnotes:

C-c C-x f
: The footnote action command.

  When the cursor is on a footnote reference, jump to the definition. When it is at a definition, jump to the (first) reference.

Otherwise, create a new footnote. Depending on the option `org-footnote-define-inline`[16], the definition will be placed right into the text as part of the reference, or separately into the location determined by the option `org-footnote-section`.

When this command is called with a prefix argument, a menu of additional options is offered:

- `s` Sort the footnote definitions by reference sequence. During editing, Org makes no effort to sort footnote definitions into a particular sequence. If you want them sorted, use this command, which will also move entries according to `org-footnote-section`. Automatic sorting after each insertion/deletion can be configured using the option `org-footnote-auto-adjust`.
- `r` Renumber the simple `fn:N` footnotes. Automatic renumbering after each insertion/deletion can be configured using the option `org-footnote-auto-adjust`.
- `S` Short for first `r`, then `s` action.
- `n` Normalize the footnotes by collecting all definitions (including inline definitions) into a special section, and then numbering them in sequence. The references will then also be numbers.
- `d` Delete the footnote at point, and all definitions of and references to it.

Depending on the variable `org-footnote-auto-adjust`[17], renumbering and sorting footnotes can be automatic after each insertion or deletion.

C-c C-c If the cursor is on a footnote reference, jump to the definition. If it is a the definition, jump back to the reference. When called at a footnote location with a prefix argument, offer the same menu as *C-c C-x f*.

C-c C-o or *mouse-1/2*
Footnote labels are also links to the corresponding definition/reference, and you can use the usual commands to follow these links.

C-c '

C-c ' Edit the footnote definition corresponding to the reference at point in a separate window. The window can be closed by pressing *C-c '*.

2.11 The Orgstruct minor mode

If you like the intuitive way the Org mode structure editing and list formatting works, you might want to use these commands in other modes like Text mode or Mail mode as well. The minor mode `orgstruct-mode` makes this possible. Toggle the mode with *M-x orgstruct-mode RET*, or turn it on by default, for example in Message mode, with one of:

```
(add-hook 'message-mode-hook 'turn-on-orgstruct)
(add-hook 'message-mode-hook 'turn-on-orgstruct++)
```

[16] The corresponding in-buffer setting is: `#+STARTUP: fninline` or `#+STARTUP: nofninline`

[17] the corresponding in-buffer options are `fnadjust` and `nofnadjust`.

When this mode is active and the cursor is on a line that looks to Org like a headline or the first line of a list item, most structure editing commands will work, even if the same keys normally have different functionality in the major mode you are using. If the cursor is not in one of those special lines, Orgstruct mode lurks silently in the shadows.

When you use `orgstruct++-mode`, Org will also export indentation and autofill settings into that mode, and detect item context after the first line of an item.

You can also use Org structure editing to fold and unfold headlines in *any* file, provided you defined `orgstruct-heading-prefix-regexp`: the regular expression must match the local prefix to use before Org's headlines. For example, if you set this variable to `";; "` in Emacs Lisp files, you will be able to fold and unfold headlines in Emacs Lisp commented lines. Some commands like `org-demote` are disabled when the prefix is set, but folding/unfolding will work correctly.

2.12 Org syntax

A reference document providing a formal description of Org's syntax is available as a draft on Worg, written and maintained by Nicolas Goaziou. It defines Org's core internal concepts such as `headlines`, `sections`, `affiliated keywords`, `(greater) elements` and `objects`. Each part of an Org file falls into one of the categories above.

To explore the abstract structure of an Org buffer, run this in a buffer:

```
M-: (org-element-parse-buffer) RET
```

It will output a list containing the buffer's content represented as an abstract structure. The export engine relies on the information stored in this list. Most interactive commands (e.g., for structure editing) also rely on the syntactic meaning of the surrounding context.

You can check syntax in your documents using `org-lint` command.

3 Tables

Org comes with a fast and intuitive table editor. Spreadsheet-like calculations are supported using the Emacs `calc` package (see *Gnu Emacs Calculator Manual*).

3.1 The built-in table editor

Org makes it easy to format tables in plain ASCII. Any line with '`|`' as the first non-whitespace character is considered part of a table. '`|`' is also the column separator[1]. A table might look like this:

```
| Name  | Phone | Age |
|-------+-------+-----|
| Peter |  1234 |  17 |
| Anna  |  4321 |  25 |
```

A table is re-aligned automatically each time you press `TAB` or `RET` or *C-c C-c* inside the table. `TAB` also moves to the next field (`RET` to the next row) and creates new table rows at the end of the table or before horizontal lines. The indentation of the table is set by the first line. Any line starting with '`|-`' is considered as a horizontal separator line and will be expanded on the next re-align to span the whole table width. So, to create the above table, you would only type

```
|Name|Phone|Age|
|-
```

and then press `TAB` to align the table and start filling in fields. Even faster would be to type `|Name|Phone|Age` followed by *C-c RET*.

When typing text into a field, Org treats `DEL`, `Backspace`, and all character keys in a special way, so that inserting and deleting avoids shifting other fields. Also, when typing *immediately after the cursor was moved into a new field with* `TAB`*, S-TAB or* `RET`, the field is automatically made blank. If this behavior is too unpredictable for you, configure the option `org-table-auto-blank-field`.

Creation and conversion

C-c | `org-table-create-or-convert-from-region`

Convert the active region to a table. If every line contains at least one TAB character, the function assumes that the material is tab separated. If every line contains a comma, comma-separated values (CSV) are assumed. If not, lines are split at whitespace into fields. You can use a prefix argument to force a specific separator: *C-u* forces CSV, *C-u C-u* forces TAB, *C-u C-u C-u* will prompt for a regular expression to match the separator, and a numeric argument N indicates that at least N consecutive spaces, or alternatively a TAB will be the separator. If there is no active region, this command creates an empty Org table. But it is easier just to start typing, like *|Name|Phone|Age RET |- TAB*.

Re-aligning and field motion

C-c C-c `org-table-align`

Re-align the table and don't move to another field.

---

[1] To insert a vertical bar into a table field, use `\vert` or, inside a word `abc\vert{}def`.

C-c SPC `org-table-blank-field`
Blank the field at point.

TAB `org-table-next-field`
Re-align the table, move to the next field. Creates a new row if necessary.

S-TAB `org-table-previous-field`
Re-align, move to previous field.

RET `org-table-next-row`
Re-align the table and move down to next row. Creates a new row if necessary. At the beginning or end of a line, RET still does NEWLINE, so it can be used to split a table.

M-a `org-table-beginning-of-field`
Move to beginning of the current table field, or on to the previous field.

M-e `org-table-end-of-field`
Move to end of the current table field, or on to the next field.

Column and row editing

M-left `org-table-move-column-left`
M-right `org-table-move-column-right`
Move the current column left/right.

M-S-left `org-table-delete-column`
Kill the current column.

M-S-right `org-table-insert-column`
Insert a new column to the left of the cursor position.

M-up `org-table-move-row-up`
M-down `org-table-move-row-down`
Move the current row up/down.

M-S-up `org-table-kill-row`
Kill the current row or horizontal line.

M-S-down `org-table-insert-row`
Insert a new row above the current row. With a prefix argument, the line is created below the current one.

C-c - `org-table-insert-hline`
Insert a horizontal line below current row. With a prefix argument, the line is created above the current line.

C-c RET `org-table-hline-and-move`
Insert a horizontal line below current row, and move the cursor into the row below that line.

C-c ^ `org-table-sort-lines`
Sort the table lines in the region. The position of point indicates the column to be used for sorting, and the range of lines is the range between the nearest horizontal separator lines, or the entire table. If point is before the first column, you will be prompted for the sorting column. If there is an active region, the

mark specifies the first line and the sorting column, while point should be in the last line to be included into the sorting. The command prompts for the sorting type (alphabetically, numerically, or by time). You can sort in normal or reverse order. You can also supply your own key extraction and comparison functions. When called with a prefix argument, alphabetic sorting will be case-sensitive.

Regions

C-c C-x M-w `org-table-copy-region`
: Copy a rectangular region from a table to a special clipboard. Point and mark determine edge fields of the rectangle. If there is no active region, copy just the current field. The process ignores horizontal separator lines.

C-c C-x C-w `org-table-cut-region`
: Copy a rectangular region from a table to a special clipboard, and blank all fields in the rectangle. So this is the "cut" operation.

C-c C-x C-y `org-table-paste-rectangle`
: Paste a rectangular region into a table. The upper left corner ends up in the current field. All involved fields will be overwritten. If the rectangle does not fit into the present table, the table is enlarged as needed. The process ignores horizontal separator lines.

M-RET `org-table-wrap-region`
: Split the current field at the cursor position and move the rest to the line below. If there is an active region, and both point and mark are in the same column, the text in the column is wrapped to minimum width for the given number of lines. A numeric prefix argument may be used to change the number of desired lines. If there is no region, but you specify a prefix argument, the current field is made blank, and the content is appended to the field above.

Calculations

C-c + `org-table-sum`
: Sum the numbers in the current column, or in the rectangle defined by the active region. The result is shown in the echo area and can be inserted with *C-y*.

S-RET `org-table-copy-down`
: When current field is empty, copy from first non-empty field above. When not empty, copy current field down to next row and move cursor along with it. Depending on the option `org-table-copy-increment`, integer field values will be incremented during copy. Integers that are too large will not be incremented. Also, a `0` prefix argument temporarily disables the increment. This key is also used by shift-selection and related modes (see Section 15.10.2 [Conflicts], page 238).

Miscellaneous

*C-c `* `org-table-edit-field`
: Edit the current field in a separate window. This is useful for fields that are not fully visible (see Section 3.2 [Column width and alignment], page 22). When called with a *C-u* prefix, just make the full field visible, so that it can be edited

in place. When called with two *C-u* prefixes, make the editor window follow the cursor through the table and always show the current field. The follow mode exits automatically when the cursor leaves the table, or when you repeat this command with *C-u C-u C-c `*.

M-x org-table-import RET
Import a file as a table. The table should be TAB or whitespace separated. Use, for example, to import a spreadsheet table or data from a database, because these programs generally can write TAB-separated text files. This command works by inserting the file into the buffer and then converting the region to a table. Any prefix argument is passed on to the converter, which uses it to determine the separator.

C-c | `org-table-create-or-convert-from-region`
Tables can also be imported by pasting tabular text into the Org buffer, selecting the pasted text with *C-x C-x* and then using the *C-c |* command (see above under *Creation and conversion*).

M-x org-table-export RET
Export the table, by default as a TAB-separated file. Use for data exchange with, for example, spreadsheet or database programs. The format used to export the file can be configured in the option `org-table-export-default-format`. You may also use properties `TABLE_EXPORT_FILE` and `TABLE_EXPORT_FORMAT` to specify the file name and the format for table export in a subtree. Org supports quite general formats for exported tables. The exporter format is the same as the format used by Orgtbl radio tables, see Section A.6.3 [Translator functions], page 246, for a detailed description.

If you don't like the automatic table editor because it gets in your way on lines which you would like to start with '`|`', you can turn it off with

```
(setq org-enable-table-editor nil)
```

Then the only table command that still works is *C-c C-c* to do a manual re-align.

3.2 Column width and alignment

The width of columns is automatically determined by the table editor. And also the alignment of a column is determined automatically from the fraction of number-like versus non-number fields in the column.

Sometimes a single field or a few fields need to carry more text, leading to inconveniently wide columns. Or maybe you want to make a table with several columns having a fixed width, regardless of content. To set the width of a column, one field anywhere in the column may contain just the string '`<N>`' where '`N`' is an integer specifying the width of the column in characters. The next re-align will then set the width of this column to this value.

```
|---+------------------------------|               |---+--------|
|   |                              |               |   | <6>    |
| 1 | one                          |               | 1 | one    |
| 2 | two                          |     ----\     | 2 | two    |
| 3 | This is a long chunk of text |     ----/     | 3 | This=> |
| 4 | four                         |               | 4 | four   |
|---+------------------------------|               |---+--------|
```

Fields that are wider become clipped and end in the string '=>'. Note that the full text is still in the buffer but is hidden. To see the full text, hold the mouse over the field—a tool-tip window will show the full content. To edit such a field, use the command *C-c `* (that is *C-c* followed by the grave accent). This will open a new window with the full field. Edit it and finish with *C-c C-c*.

When visiting a file containing a table with narrowed columns, the necessary character hiding has not yet happened, and the table needs to be aligned before it looks nice. Setting the option `org-startup-align-all-tables` will realign all tables in a file upon visiting, but also slow down startup. You can also set this option on a per-file basis with:

```
#+STARTUP: align
#+STARTUP: noalign
```

If you would like to overrule the automatic alignment of number-rich columns to the right and of string-rich columns to the left, you can use '`<r>`', '`<c>`'[2] or '`<l>`' in a similar fashion. You may also combine alignment and field width like this: '`<r10>`'.

Lines which only contain these formatting cookies will be removed automatically when exporting the document.

3.3 Column groups

When Org exports tables, it does so by default without vertical lines because that is visually more satisfying in general. Occasionally however, vertical lines can be useful to structure a table into groups of columns, much like horizontal lines can do for groups of rows. In order to specify column groups, you can use a special row where the first field contains only '/'. The further fields can either contain '<' to indicate that this column should start a group, '>' to indicate the end of a group, or '<>' (no space between '<' and '>') to make a column a group of its own. Boundaries between column groups will upon export be marked with vertical lines. Here is an example:

```
| N | N^2 | N^3 | N^4 | ~sqrt(n)~ | ~sqrt[4](N)~ |
|---+-----+-----+-----+-----------+--------------|
| / |   < |     |   > |         < |            > |
| 1 |   1 |   1 |   1 |         1 |            1 |
| 2 |   4 |   8 |  16 |    1.4142 |       1.1892 |
| 3 |   9 |  27 |  81 |    1.7321 |       1.3161 |
|---+-----+-----+-----+-----------+--------------|
#+TBLFM: $2=$1^2::$3=$1^3::$4=$1^4::$5=sqrt($1)::$6=sqrt(sqrt(($1)))
```

It is also sufficient to just insert the column group starters after every vertical line you would like to have:

[2] Centering does not work inside Emacs, but it does have an effect when exporting to HTML.

```
|  N | N^2 | N^3 | N^4 | sqrt(n) | sqrt[4](N) |
|----+-----+-----+-----+---------+------------|
| /  | <   |     |     | <       |            |
```

3.4 The Orgtbl minor mode

If you like the intuitive way the Org table editor works, you might also want to use it in other modes like Text mode or Mail mode. The minor mode Orgtbl mode makes this possible. You can always toggle the mode with *M-x orgtbl-mode RET*. To turn it on by default, for example in Message mode, use

```
(add-hook 'message-mode-hook 'turn-on-orgtbl)
```

Furthermore, with some special setup, it is possible to maintain tables in arbitrary syntax with Orgtbl mode. For example, it is possible to construct LaTeX tables with the underlying ease and power of Orgtbl mode, including spreadsheet capabilities. For details, see Section A.6 [Tables in arbitrary syntax], page 243.

3.5 The spreadsheet

The table editor makes use of the Emacs `calc` package to implement spreadsheet-like capabilities. It can also evaluate Emacs Lisp forms to derive fields from other fields. While fully featured, Org's implementation is not identical to other spreadsheets. For example, Org knows the concept of a *column formula* that will be applied to all non-header fields in a column without having to copy the formula to each relevant field. There is also a formula debugger, and a formula editor with features for highlighting fields in the table corresponding to the references at the point in the formula, moving these references by arrow keys

3.5.1 References

To compute fields in the table from other fields, formulas must reference other fields or ranges. In Org, fields can be referenced by name, by absolute coordinates, and by relative coordinates. To find out what the coordinates of a field are, press *C-c ?* in that field, or press *C-c }* to toggle the display of a grid.

Field references

Formulas can reference the value of another field in two ways. Like in any other spreadsheet, you may reference fields with a letter/number combination like `B3`, meaning the 2nd field in the 3rd row. However, Org prefers[3] to use another, more general representation that looks like this:

```
@row$column
```

Column specifications can be absolute like `$1`, `$2`,...`$N`, or relative to the current column (i.e., the column of the field which is being computed) like `$+1` or `$-2`. `$<` and `$>` are immutable references to the first and last column, respectively, and you can use `$>>>` to indicate the third column from the right.

---

[3] Org will understand references typed by the user as 'B4', but it will not use this syntax when offering a formula for editing. You can customize this behavior using the option `org-table-use-standard-references`.

The row specification only counts data lines and ignores horizontal separator lines (hlines). Like with columns, you can use absolute row numbers `@1`, `@2`,...`@`*N*, and row numbers relative to the current row like `@+3` or `@-1`. `@<` and `@>` are immutable references the first and last[4] row in the table, respectively. You may also specify the row relative to one of the hlines: `@I` refers to the first hline, `@II` to the second, etc. `@-I` refers to the first such line above the current line, `@+I` to the first such line below the current line. You can also write `@III+2` which is the second data line after the third hline in the table.

`@0` and `$0` refer to the current row and column, respectively, i.e., to the row/column for the field being computed. Also, if you omit either the column or the row part of the reference, the current row/column is implied.

Org's references with *unsigned* numbers are fixed references in the sense that if you use the same reference in the formula for two different fields, the same field will be referenced each time. Org's references with *signed* numbers are floating references because the same reference operator can reference different fields depending on the field being calculated by the formula.

Here are a few examples:

`@2$3`	2nd row, 3rd column (same as `C2`)
`$5`	column 5 in the current row (same as `E&`)
`@2`	current column, row 2
`@-1$-3`	the field one row up, three columns to the left
`@-I$2`	field just under hline above current row, column 2
`@>$5`	field in the last row, in column 5

Range references

You may reference a rectangular range of fields by specifying two field references connected by two dots '`..`'. If both fields are in the current row, you may simply use '`$2..$7`', but if at least one field is in a different row, you need to use the general `@row$column` format at least for the first field (i.e the reference must start with '`@`' in order to be interpreted correctly). Examples:

`$1..$3`	first three fields in the current row
`$P..$Q`	range, using column names (see under Advanced)
`$<<<..$>>`	start in third column, continue to the last but one
`@2$1..@4$3`	6 fields between these two fields (same as `A2..C4`)
`@-1$-2..@-1`	3 fields in the row above, starting from 2 columns on the left
`@I..II`	between first and second hline, short for `@I..@II`

Range references return a vector of values that can be fed into Calc vector functions. Empty fields in ranges are normally suppressed, so that the vector contains only the non-empty fields. For other options with the mode switches '`E`', '`N`' and examples see Section 3.5.2 [Formula syntax for Calc], page 27.

---

[4] For backward compatibility you can also use special names like `$LR5` and `$LR12` to refer in a stable way to the 5th and 12th field in the last row of the table. However, this syntax is deprecated, it should not be used for new documents. Use `@>$` instead.

Field coordinates in formulas

One of the very first actions during evaluation of Calc formulas and Lisp formulas is to substitute `@#` and `$#` in the formula with the row or column number of the field where the current result will go to. The traditional Lisp formula equivalents are `org-table-current-dline` and `org-table-current-column`. Examples:

`if(@# % 2, $#, string(""))`
: Insert column number on odd rows, set field to empty on even rows.

`$2 = '(identity remote(FOO, @@#$1))`
: Copy text or values of each row of column 1 of the table named `FOO` into column 2 of the current table.

`@3 = 2 * remote(FOO, @1$$#)`
: Insert the doubled value of each column of row 1 of the table named `FOO` into row 3 of the current table.

For the second/third example, the table named `FOO` must have at least as many rows/columns as the current table. Note that this is inefficient[5] for large number of rows/columns.

Named references

'`$name`' is interpreted as the name of a column, parameter or constant. Constants are defined globally through the option `org-table-formula-constants`, and locally (for the file) through a line like

```
#+CONSTANTS: c=299792458. pi=3.14 eps=2.4e-6
```

Also properties (see Chapter 7 [Properties and columns], page 64) can be used as constants in table formulas: for a property '`:Xyz:`' use the name '`$PROP_Xyz`', and the property will be searched in the current outline entry and in the hierarchy above it. If you have the `constants.el` package, it will also be used to resolve constants, including natural constants like '`$h`' for Planck's constant, and units like '`$km`' for kilometers[6]. Column names and parameters can be specified in special table lines. These are described below, see Section 3.5.10 [Advanced features], page 34. All names must start with a letter, and further consist of letters and numbers.

Remote references

You may also reference constants, fields and ranges from a different table, either in the current file or even in a different file. The syntax is

```
remote(NAME-OR-ID,REF)
```

where NAME can be the name of a table in the current file as set by a `#+NAME: Name` line before the table. It can also be the ID of an entry, even in a different file, and the reference then refers to the first table in that entry. REF is an absolute field or range reference as described above for example `@3$3` or `$somename`, valid in the referenced table.

---

[5] The computation time scales as O(N^2) because the table named `FOO` is parsed for each field to be read.

[6] `constants.el` can supply the values of constants in two different unit systems, `SI` and `cgs`. Which one is used depends on the value of the variable `constants-unit-system`. You can use the `#+STARTUP` options `constSI` and `constcgs` to set this value for the current buffer.

Indirection of NAME-OR-ID: When NAME-OR-ID has the format `@ROW$COLUMN` it will be substituted with the name or ID found in this field of the current table. For example `remote($1, @>$2)` => `remote(year_2013, @>$1)`. The format `B3` is not supported because it can not be distinguished from a plain table name or ID.

3.5.2 Formula syntax for Calc

A formula can be any algebraic expression understood by the Emacs `Calc` package. Note that `calc` has the non-standard convention that '`/`' has lower precedence than '`*`', so that '`a/b*c`' is interpreted as '`a/(b*c)`'. Before evaluation by `calc-eval` (see Section "Calling Calc from Your Lisp Programs" in *GNU Emacs Calc Manual*), variable substitution takes place according to the rules described above. The range vectors can be directly fed into the Calc vector functions like '`vmean`' and '`vsum`'.

A formula can contain an optional mode string after a semicolon. This string consists of flags to influence Calc and other modes during execution. By default, Org uses the standard Calc modes (precision 12, angular units degrees, fraction and symbolic modes off). The display format, however, has been changed to `(float 8)` to keep tables compact. The default settings can be configured using the option `org-calc-default-modes`.

List of modes:

`p20` Set the internal Calc calculation precision to 20 digits.

`n3`, `s3`, `e2`, `f4`
Normal, scientific, engineering or fixed format of the result of Calc passed back to Org. Calc formatting is unlimited in precision as long as the Calc calculation precision is greater.

`D`, `R` Degree and radian angle modes of Calc.

`F`, `S` Fraction and symbolic modes of Calc.

`T`, `t`, `U` Duration computations in Calc or Lisp, see Section 3.5.4 [Durations and time values], page 29.

`E` If and how to consider empty fields. Without '`E`' empty fields in range references are suppressed so that the Calc vector or Lisp list contains only the non-empty fields. With '`E`' the empty fields are kept. For empty fields in ranges or empty field references the value '`nan`' (not a number) is used in Calc formulas and the empty string is used for Lisp formulas. Add '`N`' to use 0 instead for both formula types. For the value of a field the mode '`N`' has higher precedence than '`E`'.

`N` Interpret all fields as numbers, use 0 for non-numbers. See the next section to see how this is essential for computations with Lisp formulas. In Calc formulas it is used only occasionally because there number strings are already interpreted as numbers without '`N`'.

`L` Literal, for Lisp formulas only. See the next section.

Unless you use large integer numbers or high-precision-calculation and -display for floating point numbers you may alternatively provide a '`printf`' format specifier to reformat the

Calc result after it has been passed back to Org instead of letting Calc already do the formatting[7]. A few examples:

`$1+$2`	Sum of first and second field
`$1+$2;%.2f`	Same, format result to two decimals
`exp($2)+exp($1)`	Math functions can be used
`$0;%.1f`	Reformat current cell to 1 decimal
`($3-32)*5/9`	Degrees F -> C conversion
`$c/$1/$cm`	Hz -> cm conversion, using `constants.el`
`tan($1);Dp3s1`	Compute in degrees, precision 3, display SCI 1
`sin($1);Dp3%.1e`	Same, but use printf specifier for display
`taylor($3,x=7,2)`	Taylor series of $3, at x=7, second degree

Calc also contains a complete set of logical operations, (see Section "Logical Operations" in *GNU Emacs Calc Manual*). For example

`if($1 < 20, teen, string(""))`
: "teen" if age $1 is less than 20, else the Org table result field is set to empty with the empty string.

`if("$1" == "nan" || "$2" == "nan", string(""), $1 + $2); E f-1`
: Sum of the first two columns. When at least one of the input fields is empty the Org table result field is set to empty. 'E' is required to not convert empty fields to 0. 'f-1' is an optional Calc format string similar to '%.1f' but leaves empty results empty.

`if(typeof(vmean($1..$7)) == 12, string(""), vmean($1..$7); E`
: Mean value of a range unless there is any empty field. Every field in the range that is empty is replaced by 'nan' which lets 'vmean' result in 'nan'. Then 'typeof == 12' detects the 'nan' from 'vmean' and the Org table result field is set to empty. Use this when the sample set is expected to never have missing values.

`if("$1..$7" == "[]", string(""), vmean($1..$7))`
: Mean value of a range with empty fields skipped. Every field in the range that is empty is skipped. When all fields in the range are empty the mean value is not defined and the Org table result field is set to empty. Use this when the sample set can have a variable size.

`vmean($1..$7); EN`
: To complete the example before: Mean value of a range with empty fields counting as samples with value 0. Use this only when incomplete sample sets should be padded with 0 to the full size.

You can add your own Calc functions defined in Emacs Lisp with `defmath` and use them in formula syntax for Calc.

---

[7] The '`printf`' reformatting is limited in precision because the value passed to it is converted into an '`integer`' or '`double`'. The '`integer`' is limited in size by truncating the signed value to 32 bits. The '`double`' is limited in precision to 64 bits overall which leaves approximately 16 significant decimal digits.

3.5.3 Emacs Lisp forms as formulas

It is also possible to write a formula in Emacs Lisp. This can be useful for string manipulation and control structures, if Calc's functionality is not enough.

If a formula starts with an apostrophe followed by an opening parenthesis, then it is evaluated as a Lisp form. The evaluation should return either a string or a number. Just as with `calc` formulas, you can specify modes and a printf format after a semicolon.

With Emacs Lisp forms, you need to be conscious about the way field references are interpolated into the form. By default, a reference will be interpolated as a Lisp string (in double-quotes) containing the field. If you provide the 'N' mode switch, all referenced elements will be numbers (non-number fields will be zero) and interpolated as Lisp numbers, without quotes. If you provide the 'L' flag, all fields will be interpolated literally, without quotes. I.e., if you want a reference to be interpreted as a string by the Lisp form, enclose the reference operator itself in double-quotes, like `"$3"`. Ranges are inserted as space-separated fields, so you can embed them in list or vector syntax.

Here are a few examples—note how the 'N' mode is used when we do computations in Lisp:

`'(concat (substring $1 1 2) (substring $1 0 1) (substring $1 2))`
: Swap the first two characters of the content of column 1.

`'(+ $1 $2);N`
: Add columns 1 and 2, equivalent to Calc's `$1+$2`.

`'(apply '+ '($1..$4));N`
: Compute the sum of columns 1 to 4, like Calc's `vsum($1..$4)`.

3.5.4 Durations and time values

If you want to compute time values use the `T`, `t`, or `U` flag, either in Calc formulas or Elisp formulas:

```
| Task 1  |  Task 2  |   Total  |
|---------+----------+----------|
|    2:12 |     1:47 | 03:59:00 |
|    2:12 |     1:47 |    03:59 |
| 3:02:20 | -2:07:00 |     0.92 |
#+TBLFM: @2$3=$1+$2;T::@3$3=$1+$2;U::@4$3=$1+$2;t
```

Input duration values must be of the form `HH:MM[:SS]`, where seconds are optional. With the `T` flag, computed durations will be displayed as `HH:MM:SS` (see the first formula above). With the `U` flag, seconds will be omitted so that the result will be only `HH:MM` (see second formula above). Zero-padding of the hours field will depend upon the value of the variable `org-table-duration-hour-zero-padding`.

With the `t` flag, computed durations will be displayed according to the value of the option `org-table-duration-custom-format`, which defaults to `'hours` and will display the result as a fraction of hours (see the third formula in the example above).

Negative duration values can be manipulated as well, and integers will be considered as seconds in addition and subtraction.

3.5.5 Field and range formulas

To assign a formula to a particular field, type it directly into the field, preceded by ':=', for example ':=vsum(@II..III)'. When you press TAB or RET or *C-c C-c* with the cursor still in the field, the formula will be stored as the formula for this field, evaluated, and the current field will be replaced with the result.

Formulas are stored in a special line starting with '#+TBLFM:' directly below the table. If you type the equation in the 4th field of the 3rd data line in the table, the formula will look like '@3$4=$1+$2'. When inserting/deleting/swapping columns and rows with the appropriate commands, *absolute references* (but not relative ones) in stored formulas are modified in order to still reference the same field. To avoid this, in particular in range references, anchor ranges at the table borders (using @<, @>, $<, $>), or at hlines using the @I notation. Automatic adaptation of field references does of course not happen if you edit the table structure with normal editing commands—then you must fix the equations yourself.

Instead of typing an equation into the field, you may also use the following command

C-u C-c = **org-table-eval-formula**
: Install a new formula for the current field. The command prompts for a formula with default taken from the '#+TBLFM:' line, applies it to the current field, and stores it.

The left-hand side of a formula can also be a special expression in order to assign the formula to a number of different fields. There is no keyboard shortcut to enter such range formulas. To add them, use the formula editor (see Section 3.5.8 [Editing and debugging formulas], page 31) or edit the #+TBLFM: line directly.

$2=
: Column formula, valid for the entire column. This is so common that Org treats these formulas in a special way, see Section 3.5.6 [Column formulas], page 30.

@3=
: Row formula, applies to all fields in the specified row. @>= means the last row.

@1$2..@4$3=
: Range formula, applies to all fields in the given rectangular range. This can also be used to assign a formula to some but not all fields in a row.

$name=
: Named field, see Section 3.5.10 [Advanced features], page 34.

3.5.6 Column formulas

When you assign a formula to a simple column reference like $3=, the same formula will be used in all fields of that column, with the following very convenient exceptions: (i) If the table contains horizontal separator hlines with rows above and below, everything before the first such hline is considered part of the table *header* and will not be modified by column formulas. Therefore a header is mandatory when you use column formulas and want to add hlines to group rows, like for example to separate a total row at the bottom from the summand rows above. (ii) Fields that already get a value from a field/range formula will be left alone by column formulas. These conditions make column formulas very easy to use.

To assign a formula to a column, type it directly into any field in the column, preceded by an equal sign, like '=$1+$2'. When you press TAB or RET or *C-c C-c* with the cursor still in the field, the formula will be stored as the formula for the current column, evaluated and the current field replaced with the result. If the field contains only '=', the previously stored formula for this column is used. For each column, Org will only remember the most recently used formula. In the '#+TBLFM:' line, column formulas will look like '$4=$1+$2'. The left-hand side of a column formula cannot be the name of column, it must be the numeric column reference or $>.

Instead of typing an equation into the field, you may also use the following command:

C-c = `org-table-eval-formula`
: Install a new formula for the current column and replace current field with the result of the formula. The command prompts for a formula, with default taken from the '#+TBLFM' line, applies it to the current field and stores it. With a numeric prefix argument(e.g., *C-5 C-c =*) the command will apply it to that many consecutive fields in the current column.

3.5.7 Lookup functions

Org has three predefined Emacs Lisp functions for lookups in tables.

`(org-lookup-first VAL S-LIST R-LIST &optional PREDICATE)`
: Searches for the first element `S` in list `S-LIST` for which

    `(PREDICATE VAL S)`

  is `t`; returns the value from the corresponding position in list `R-LIST`. The default `PREDICATE` is `equal`. Note that the parameters `VAL` and `S` are passed to `PREDICATE` in the same order as the corresponding parameters are in the call to `org-lookup-first`, where `VAL` precedes `S-LIST`. If `R-LIST` is `nil`, the matching element `S` of `S-LIST` is returned.

`(org-lookup-last VAL S-LIST R-LIST &optional PREDICATE)`
: Similar to `org-lookup-first` above, but searches for the *last* element for which `PREDICATE` is `t`.

`(org-lookup-all VAL S-LIST R-LIST &optional PREDICATE)`
: Similar to `org-lookup-first`, but searches for *all* elements for which `PREDICATE` is `t`, and returns *all* corresponding values. This function can not be used by itself in a formula, because it returns a list of values. However, powerful lookups can be built when this function is combined with other Emacs Lisp functions.

If the ranges used in these functions contain empty fields, the `E` mode for the formula should usually be specified: otherwise empty fields will not be included in `S-LIST` and/or `R-LIST` which can, for example, result in an incorrect mapping from an element of `S-LIST` to the corresponding element of `R-LIST`.

These three functions can be used to implement associative arrays, count matching cells, rank results, group data etc. For practical examples see this tutorial on Worg.

3.5.8 Editing and debugging formulas

You can edit individual formulas in the minibuffer or directly in the field. Org can also prepare a special buffer with all active formulas of a table. When offering a formula for editing, Org converts references to the standard format (like `B3` or `D&`) if possible. If you prefer to only work with the internal format (like `@3$2` or `$4`), configure the option `org-table-use-standard-references`.

C-c = or *C-u C-c =* `org-table-eval-formula`
: Edit the formula associated with the current column/field in the minibuffer. See Section 3.5.6 [Column formulas], page 30, and Section 3.5.5 [Field and range formulas], page 30.

C-u C-u C-c = `org-table-eval-formula`
: Re-insert the active formula (either a field formula, or a column formula) into the current field, so that you can edit it directly in the field. The advantage over editing in the minibuffer is that you can use the command *C-c ?*.

C-c ? `org-table-field-info`
: While editing a formula in a table field, highlight the field(s) referenced by the reference at the cursor position in the formula.

C-c } Toggle the display of row and column numbers for a table, using overlays (`org-table-toggle-coordinate-overlays`). These are updated each time the table is aligned; you can force it with *C-c C-c*.

C-c { Toggle the formula debugger on and off (`org-table-toggle-formula-debugger`). See below.

C-c ' `org-table-edit-formulas`
: Edit all formulas for the current table in a special buffer, where the formulas will be displayed one per line. If the current field has an active formula, the cursor in the formula editor will mark it. While inside the special buffer, Org will automatically highlight any field or range reference at the cursor position. You may edit, remove and add formulas, and use the following commands:

  C-c C-c or *C-x C-s* `org-table-fedit-finish`
  : Exit the formula editor and store the modified formulas. With *C-u* prefix, also apply the new formulas to the entire table.

  C-c C-q `org-table-fedit-abort`
  : Exit the formula editor without installing changes.

  C-c C-r `org-table-fedit-toggle-ref-type`
  : Toggle all references in the formula editor between standard (like `B3`) and internal (like `@3$2`).

  TAB `org-table-fedit-lisp-indent`
  : Pretty-print or indent Lisp formula at point. When in a line containing a Lisp formula, format the formula according to Emacs Lisp rules. Another `TAB` collapses the formula back again. In the open formula, `TAB` re-indents just like in Emacs Lisp mode.

  M-TAB `lisp-complete-symbol`
  : Complete Lisp symbols, just like in Emacs Lisp mode.[8]

---

[8] Many desktops intercept *M-TAB* to switch windows. Use *C-M-i* or *ESC TAB* instead for completion (see Section 15.1 [Completion], page 228).

S-up/down/left/right
: Shift the reference at point. For example, if the reference is `B3` and you press *S-right*, it will become `C3`. This also works for relative references and for hline references.

M-S-up `org-table-fedit-line-up`
M-S-down `org-table-fedit-line-down`
: Move the test line for column formulas in the Org buffer up and down.

M-up `org-table-fedit-scroll-down`
M-down `org-table-fedit-scroll-up`
: Scroll the window displaying the table.

C-c } Turn the coordinate grid in the table on and off.

Making a table field blank does not remove the formula associated with the field, because that is stored in a different line (the '`#+TBLFM`' line)—during the next recalculation the field will be filled again. To remove a formula from a field, you have to give an empty reply when prompted for the formula, or to edit the '`#+TBLFM`' line.

You may edit the '`#+TBLFM`' directly and re-apply the changed equations with *C-c C-c* in that line or with the normal recalculation commands in the table.

Using multiple #+TBLFM lines

You may apply the formula temporarily. This is useful when you switch the formula. Place multiple '`#+TBLFM`' lines right after the table, and then press *C-c C-c* on the formula to apply. Here is an example:

```
| x | y |
|---+---|
| 1 |   |
| 2 |   |
#+TBLFM: $2=$1*1
#+TBLFM: $2=$1*2
```

Pressing *C-c C-c* in the line of '`#+TBLFM: $2=$1*2`' yields:

```
| x | y |
|---+---|
| 1 | 2 |
| 2 | 4 |
#+TBLFM: $2=$1*1
#+TBLFM: $2=$1*2
```

Note: If you recalculate this table (with *C-u C-c **, for example), you will get the following result of applying only the first '`#+TBLFM`' line.

```
| x | y |
|---+---|
| 1 | 1 |
| 2 | 2 |
#+TBLFM: $2=$1*1
#+TBLFM: $2=$1*2
```

Debugging formulas

When the evaluation of a formula leads to an error, the field content becomes the string '#ERROR'. If you would like see what is going on during variable substitution and calculation in order to find a bug, turn on formula debugging in the Tbl menu and repeat the calculation, for example by pressing *C-u C-u C-c = RET* in a field. Detailed information will be displayed.

3.5.9 Updating the table

Recalculation of a table is normally not automatic, but needs to be triggered by a command. See Section 3.5.10 [Advanced features], page 34, for a way to make recalculation at least semi-automatic.

In order to recalculate a line of a table or the entire table, use the following commands:

*C-c ** `org-table-recalculate`
: Recalculate the current row by first applying the stored column formulas from left to right, and all field/range formulas in the current row.

*C-u C-c **
C-u C-c C-c
: Recompute the entire table, line by line. Any lines before the first hline are left alone, assuming that these are part of the table header.

*C-u C-u C-c ** or *C-u C-u C-c C-c* `org-table-iterate`
: Iterate the table by recomputing it until no further changes occur. This may be necessary if some computed fields use the value of other fields that are computed *later* in the calculation sequence.

M-x org-table-recalculate-buffer-tables RET
: Recompute all tables in the current buffer.

M-x org-table-iterate-buffer-tables RET
: Iterate all tables in the current buffer, in order to converge table-to-table dependencies.

3.5.10 Advanced features

If you want the recalculation of fields to happen automatically, or if you want to be able to assign *names*[9] to fields and columns, you need to reserve the first column of the table for special marking characters.

C-# `org-table-rotate-recalc-marks`
: Rotate the calculation mark in first column through the states ' ', '#', '*', '!', '$'. When there is an active region, change all marks in the region.

Here is an example of a table that collects exam results of students and makes use of these features:

---

[9] Such names must start by an alphabetic character and use only alphanumeric/underscore characters.

```
|---+---------+--------+--------+--------+-------+------|
|   | Student | Prob 1 | Prob 2 | Prob 3 | Total | Note |
|---+---------+--------+--------+--------+-------+------|
| ! |         |     P1 |     P2 |     P3 |   Tot |      |
| # | Maximum |     10 |     15 |     25 |    50 | 10.0 |
| ^ |         |     m1 |     m2 |     m3 |    mt |      |
|---+---------+--------+--------+--------+-------+------|
| # | Peter   |     10 |      8 |     23 |    41 |  8.2 |
| # | Sam     |      2 |      4 |      3 |     9 |  1.8 |
|---+---------+--------+--------+--------+-------+------|
|   | Average |        |        |        |  25.0 |      |
| ^ |         |        |        |        |    at |      |
| $ | max=50  |        |        |        |       |      |
|---+---------+--------+--------+--------+-------+------|
#+TBLFM: $6=vsum($P1..$P3)::$7=10*$Tot/$max;%.1f::$at=vmean(@-II..@-I);%.1f
```

Important: please note that for these special tables, recalculating the table with *C-u C-c ** will only affect rows that are marked '#' or '*', and fields that have a formula assigned to the field itself. The column formulas are not applied in rows with empty first field.

The marking characters have the following meaning:

'!' The fields in this line define names for the columns, so that you may refer to a column as '$Tot' instead of '$6'.

'^' This row defines names for the fields *above* the row. With such a definition, any formula in the table may use '$m1' to refer to the value '10'. Also, if you assign a formula to a names field, it will be stored as '$name=...'.

'_' Similar to '^', but defines names for the fields in the row *below*.

'$' Fields in this row can define *parameters* for formulas. For example, if a field in a '$' row contains 'max=50', then formulas in this table can refer to the value 50 using '$max'. Parameters work exactly like constants, only that they can be defined on a per-table basis.

'#' Fields in this row are automatically recalculated when pressing TAB or RET or *S-TAB* in this row. Also, this row is selected for a global recalculation with *C-u C-c **. Unmarked lines will be left alone by this command.

'*' Selects this line for global recalculation with *C-u C-c **, but not for automatic recalculation. Use this when automatic recalculation slows down editing too much.

' ' Unmarked lines are exempt from recalculation with *C-u C-c **. All lines that should be recalculated should be marked with '#' or '*'.

'/' Do not export this line. Useful for lines that contain the narrowing '<N>' markers or column group markers.

Finally, just to whet your appetite for what can be done with the fantastic calc.el package, here is a table that computes the Taylor series of degree n at location x for a couple of functions.

```
|---+-------------+---+-----+--------------------------------------|
|   | Func        | n | x   | Result                               |
|---+-------------+---+-----+--------------------------------------|
| # | exp(x)      | 1 | x   | 1 + x                                |
| # | exp(x)      | 2 | x   | 1 + x + x^2 / 2                      |
| # | exp(x)      | 3 | x   | 1 + x + x^2 / 2 + x^3 / 6            |
| # | x^2+sqrt(x) | 2 | x=0 | x*(0.5 / 0) + x^2 (2 - 0.25 / 0) / 2 |
| # | x^2+sqrt(x) | 2 | x=1 | 2 + 2.5 x - 2.5 + 0.875 (x - 1)^2    |
| * | tan(x)      | 3 | x   | 0.0175 x + 1.77e-6 x^3               |
|---+-------------+---+-----+--------------------------------------|
#+TBLFM: $5=taylor($2,$4,$3);n3
```

3.6 Org-Plot

Org-Plot can produce graphs of information stored in org tables, either graphically or in ASCII-art.

Graphical plots using `Gnuplot`

Org-Plot produces 2D and 3D graphs using `Gnuplot` `http://www.gnuplot.info/` and `gnuplot-mode` `http://xafs.org/BruceRavel/GnuplotMode`. To see this in action, ensure that you have both Gnuplot and Gnuplot mode installed on your system, then call *C-c " g* or *M-x org-plot/gnuplot RET* on the following table.

```
#+PLOT: title:"Citas" ind:1 deps:(3) type:2d with:histograms set:"yrange [0:]"
| Sede       | Max cites | H-index |
|------------+-----------+---------|
| Chile      |    257.72 |   21.39 |
| Leeds      |    165.77 |   19.68 |
| Sao Paolo  |     71.00 |   11.50 |
| Stockholm  |    134.19 |   14.33 |
| Morelia    |    257.56 |   17.67 |
```

Notice that Org Plot is smart enough to apply the table's headers as labels. Further control over the labels, type, content, and appearance of plots can be exercised through the `#+PLOT:` lines preceding a table. See below for a complete list of Org-plot options. The `#+PLOT:` lines are optional. For more information and examples see the Org-plot tutorial at `http://orgmode.org/worg/org-tutorials/org-plot.html`.

Plot Options

`set` Specify any `gnuplot` option to be set when graphing.

`title` Specify the title of the plot.

`ind` Specify which column of the table to use as the `x` axis.

`deps` Specify the columns to graph as a Lisp style list, surrounded by parentheses and separated by spaces for example `dep:(3 4)` to graph the third and fourth columns (defaults to graphing all other columns aside from the `ind` column).

`type` Specify whether the plot will be `2d`, `3d`, or `grid`.

`with` Specify a `with` option to be inserted for every col being plotted (e.g., `lines`, `points`, `boxes`, `impulses`, etc...). Defaults to `lines`.

`file` If you want to plot to a file, specify `"path/to/desired/output-file"`.

`labels` List of labels to be used for the `deps` (defaults to the column headers if they exist).

`line` Specify an entire line to be inserted in the Gnuplot script.

`map` When plotting `3d` or `grid` types, set this to `t` to graph a flat mapping rather than a `3d` slope.

`timefmt` Specify format of Org mode timestamps as they will be parsed by Gnuplot. Defaults to '`%Y-%m-%d-%H:%M:%S`'.

`script` If you want total control, you can specify a script file (place the file name between double-quotes) which will be used to plot. Before plotting, every instance of `$datafile` in the specified script will be replaced with the path to the generated data file. Note: even if you set this option, you may still want to specify the plot type, as that can impact the content of the data file.

ASCII bar plots

While the cursor is on a column, typing `C-c " a` or `M-x orgtbl-ascii-plot RET` create a new column containing an ASCII-art bars plot. The plot is implemented through a regular column formula. When the source column changes, the bar plot may be updated by refreshing the table, for example typing `C-u C-c *`.

```
| Sede          | Max cites |              |
|---------------+-----------+--------------|
| Chile         |    257.72 | WWWWWWWWWWWW |
| Leeds         |    165.77 | WWWWWWWh     |
| Sao Paolo     |     71.00 | WWW;         |
| Stockholm     |    134.19 | WWWWWW:      |
| Morelia       |    257.56 | WWWWWWWWWWWH |
| Rochefourchat |      0.00 |              |
#+TBLFM: $3='(orgtbl-ascii-draw $2 0.0 257.72 12)
```

The formula is an elisp call:

```
(orgtbl-ascii-draw COLUMN MIN MAX WIDTH)
```

`COLUMN` is a reference to the source column.

`MIN MAX` are the minimal and maximal values displayed. Sources values outside this range are displayed as '`too small`' or '`too large`'.

`WIDTH` is the width in characters of the bar-plot. It defaults to '`12`'.

4 Hyperlinks

Like HTML, Org provides links inside a file, external links to other files, Usenet articles, emails, and much more.

4.1 Link format

Org will recognize plain URL-like links and activate them as clickable links. The general link format, however, looks like this:

`[[link][description]]`   or alternatively   `[[link]]`

Once a link in the buffer is complete (all brackets present), Org will change the display so that '`description`' is displayed instead of '`[[link][description]]`' and '`link`' is displayed instead of '`[[link]]`'. Links will be highlighted in the face `org-link`, which by default is an underlined face. You can directly edit the visible part of a link. Note that this can be either the '`link`' part (if there is no description) or the '`description`' part. To edit also the invisible '`link`' part, use *C-c C-l* with the cursor on the link.

If you place the cursor at the beginning or just behind the end of the displayed text and press `BACKSPACE`, you will remove the (invisible) bracket at that location. This makes the link incomplete and the internals are again displayed as plain text. Inserting the missing bracket hides the link internals again. To show the internal structure of all links, use the menu entry `Org->Hyperlinks->Literal links`.

4.2 Internal links

If the link does not look like a URL, it is considered to be internal in the current file. The most important case is a link like '`[[#my-custom-id]]`' which will link to the entry with the `CUSTOM_ID` property '`my-custom-id`'. You are responsible yourself to make sure these custom IDs are unique in a file.

Links such as '`[[My Target]]`' or '`[[My Target][Find my target]]`' lead to a text search in the current file.

The link can be followed with *C-c C-o* when the cursor is on the link, or with a mouse click (see Section 4.4 [Handling links], page 41). Links to custom IDs will point to the corresponding headline. The preferred match for a text link is a *dedicated target*: the same string in double angular brackets, like '<<My Target>>'.

If no dedicated target exists, the link will then try to match the exact name of an element within the buffer. Naming is done with the `#+NAME` keyword, which has to be put in the line before the element it refers to, as in the following example

```
#+NAME: My Target
| a  | table      |
|----+------------|
| of | four cells |
```

If none of the above succeeds, Org will search for a headline that is exactly the link text but may also include a TODO keyword and tags[1].

---

[1] To insert a link targeting a headline, in-buffer completion can be used. Just type a star followed by a few optional letters into the buffer and press *M-TAB*. All headlines in the current buffer will be offered as completions.

During export, internal links will be used to mark objects and assign them a number. Marked objects will then be referenced by links pointing to them. In particular, links without a description will appear as the number assigned to the marked object[2]. In the following excerpt from an Org buffer

```
- one item
- <<target>>another item
Here we refer to item [[target]].
```

The last sentence will appear as 'Here we refer to item 2' when exported.

In non-Org files, the search will look for the words in the link text. In the above example the search would be for 'my target'.

Following a link pushes a mark onto Org's own mark ring. You can return to the previous position with *C-c &*. Using this command several times in direct succession goes back to positions recorded earlier.

4.2.1 Radio targets

Org can automatically turn any occurrences of certain target names in normal text into a link. So without explicitly creating a link, the text connects to the target radioing its position. Radio targets are enclosed by triple angular brackets. For example, a target '<<<My Target>>>' causes each occurrence of 'my target' in normal text to become activated as a link. The Org file is scanned automatically for radio targets only when the file is first loaded into Emacs. To update the target list during editing, press *C-c C-c* with the cursor on or at a target.

4.3 External links

Org supports links to files, websites, Usenet and email messages, BBDB database entries and links to both IRC conversations and their logs. External links are URL-like locators. They start with a short identifying string followed by a colon. There can be no space after the colon. The following list shows examples for each link type.

`http://www.astro.uva.nl/~dominik`	on the web
`doi:10.1000/182`	DOI for an electronic resource
`file:/home/dominik/images/jupiter.jpg`	file, absolute path
`/home/dominik/images/jupiter.jpg`	same as above
`file:papers/last.pdf`	file, relative path
`./papers/last.pdf`	same as above
`file:/ssh:myself@some.where:papers/last.pdf`	file, path on remote machine
`/ssh:myself@some.where:papers/last.pdf`	same as above
`file:sometextfile::NNN`	file, jump to line number
`file:projects.org`	another Org file
`file:projects.org::some words`	text search in Org file[3]

---

[2] When targeting a #+NAME keyword, #+CAPTION keyword is mandatory in order to get proper numbering (see Section 11.4 [Images and tables], page 131).

[3] The actual behavior of the search will depend on the value of the option org-link-search-must-match-exact-headline. If its value is nil, then a fuzzy text search will be done. If it is t, then only

`file:projects.org::*task title`	heading search in Org file[4]
`docview:papers/last.pdf::NNN`	open in doc-view mode at page
`id:B7423F4D-2E8A-471B-8810-C40F074717E9`	Link to heading by ID
`news:comp.emacs`	Usenet link
`mailto:adent@galaxy.net`	Mail link
`mhe:folder`	MH-E folder link
`mhe:folder#id`	MH-E message link
`rmail:folder`	RMAIL folder link
`rmail:folder#id`	RMAIL message link
`gnus:group`	Gnus group link
`gnus:group#id`	Gnus article link
`bbdb:R.*Stallman`	BBDB link (with regexp)
`irc:/irc.com/#emacs/bob`	IRC link
`info:org#External links`	Info node or index link
`shell:ls *.org`	A shell command
`elisp:org-agenda`	Interactive Elisp command
`elisp:(find-file-other-frame "Elisp.org")`	Elisp form to evaluate

On top of these built-in link types, some are available through the `contrib/` directory (see Section 1.2 [Installation], page 2). For example, these links to VM or Wanderlust messages are available when you load the corresponding libraries from the `contrib/` directory:

`vm:folder`	VM folder link
`vm:folder#id`	VM message link
`vm://myself@some.where.org/folder#id`	VM on remote machine
`vm-imap:account:folder`	VM IMAP folder link
`vm-imap:account:folder#id`	VM IMAP message link
`wl:folder`	WANDERLUST folder link
`wl:folder#id`	WANDERLUST message link

For customizing Org to add new link types Section A.3 [Adding hyperlink types], page 241.

A link should be enclosed in double brackets and may contain a descriptive text to be displayed instead of the URL (see Section 4.1 [Link format], page 38), for example:

```
[[https://www.gnu.org/software/emacs/][GNU Emacs]]
```

If the description is a file name or URL that points to an image, HTML export (see Section 12.9 [HTML export], page 149) will inline the image as a clickable button. If there is no description at all and the link points to an image, that image will be inlined into the exported HTML file.

---

the exact headline will be matched, ignoring spaces and cookies. If the value is `query-to-create`, then an exact headline will be searched; if it is not found, then the user will be queried to create it.

[4] Headline searches always match the exact headline, ignoring spaces and cookies. If the headline is not found and the value of the option `org-link-search-must-match-exact-headline` is `query-to-create`, then the user will be queried to create it.

Org also finds external links in the normal text and activates them as links. If spaces must be part of the link (for example in '`bbdb:Richard Stallman`'), or if you need to remove ambiguities about the end of the link, enclose them in square brackets.

4.4 Handling links

Org provides methods to create a link in the correct syntax, to insert it into an Org file, and to follow the link.

C-c l `org-store-link`

Store a link to the current location. This is a *global* command (you must create the key binding yourself) which can be used in any buffer to create a link. The link will be stored for later insertion into an Org buffer (see below). What kind of link will be created depends on the current buffer:

Org mode buffers
For Org files, if there is a '`<<target>>`' at the cursor, the link points to the target. Otherwise it points to the current headline, which will also be the description[5].

If the headline has a `CUSTOM_ID` property, a link to this custom ID will be stored. In addition or alternatively (depending on the value of `org-id-link-to-org-use-id`), a globally unique `ID` property will be created and/or used to construct a link[6]. So using this command in Org buffers will potentially create two links: a human-readable from the custom ID, and one that is globally unique and works even if the entry is moved from file to file. Later, when inserting the link, you need to decide which one to use.

Email/News clients: VM, Rmail, Wanderlust, MH-E, Gnus
Pretty much all Emacs mail clients are supported. The link will point to the current article, or, in some GNUS buffers, to the group. The description is constructed from the author and the subject.

Web browsers: Eww, W3 and W3M
Here the link will be the current URL, with the page title as description.

Contacts: BBDB
Links created in a BBDB buffer will point to the current entry.

Chat: IRC
For IRC links, if you set the option `org-irc-link-to-logs` to `t`, a '`file:/`' style link to the relevant point in the logs for the current conversation is created. Otherwise an '`irc:/`' style link to the user/channel/server under the point will be stored.

Other files
For any other files, the link will point to the file, with a search string (see Section 4.7 [Search options], page 45) pointing to the contents of the current line. If there is an active region, the selected words will form the basis of the search

---

[5] If the headline contains a timestamp, it will be removed from the link and result in a wrong link—you should avoid putting timestamp in the headline.

[6] The library `org-id.el` must first be loaded, either through `org-customize` by enabling `org-id` in `org-modules`, or by adding `(require 'org-id)` in your Emacs init file.

string. If the automatically created link is not working correctly or accurately enough, you can write custom functions to select the search string and to do the search for particular file types—see Section 4.8 [Custom searches], page 45. The key binding *C-c l* is only a suggestion—see Section 1.2 [Installation], page 2.

Agenda view
When the cursor is in an agenda view, the created link points to the entry referenced by the current line.

C-c C-l `org-insert-link`

Insert a link[7]. This prompts for a link to be inserted into the buffer. You can just type a link, using text for an internal link, or one of the link type prefixes mentioned in the examples above. The link will be inserted into the buffer[8], along with a descriptive text. If some text was selected when this command is called, the selected text becomes the default description.

Inserting stored links
All links stored during the current session are part of the history for this prompt, so you can access them with `up` and `down` (or *M-p/n*).

Completion support
Completion with `TAB` will help you to insert valid link prefixes like '`https:`', including the prefixes defined through link abbreviations (see Section 4.6 [Link abbreviations], page 44). If you press `RET` after inserting only the *prefix*, Org will offer specific completion support for some link types[9] For example, if you type *file RET*, file name completion (alternative access: *C-u C-c C-l*, see below) will be offered, and after *bbdb RET* you can complete contact names.

C-u C-c C-l

When *C-c C-l* is called with a *C-u* prefix argument, a link to a file will be inserted and you may use file name completion to select the name of the file. The path to the file is inserted relative to the directory of the current Org file, if the linked file is in the current directory or in a sub-directory of it, or if the path is written relative to the current directory using '`../`'. Otherwise an absolute path is used, if possible with '`~/`' for your home directory. You can force an absolute path with two *C-u* prefixes.

C-c C-l (with cursor on existing link)

When the cursor is on an existing link, *C-c C-l* allows you to edit the link and description parts of the link.

C-c C-o `org-open-at-point`

Open link at point. This will launch a web browser for URLs (using `browse-url-at-point`), run VM/MH-E/Wanderlust/Rmail/Gnus/BBDB for

---

[7] Note that you don't have to use this command to insert a link. Links in Org are plain text, and you can type or paste them straight into the buffer. By using this command, the links are automatically enclosed in double brackets, and you will be asked for the optional descriptive text.

[8] After insertion of a stored link, the link will be removed from the list of stored links. To keep it in the list later use, use a triple *C-u* prefix argument to *C-c C-l*, or configure the option `org-keep-stored-link-after-insertion`.

[9] This works if a completion function is defined in the '`:complete`' property of a link in `org-link-parameters`.

the corresponding links, and execute the command in a shell link. When the cursor is on an internal link, this command runs the corresponding search. When the cursor is on a TAG list in a headline, it creates the corresponding TAGS view. If the cursor is on a timestamp, it compiles the agenda for that date. Furthermore, it will visit text and remote files in 'file:' links with Emacs and select a suitable application for local non-text files. Classification of files is based on file extension only. See option `org-file-apps`. If you want to override the default application and visit the file with Emacs, use a *C-u* prefix. If you want to avoid opening in Emacs, use a *C-u C-u* prefix.
If the cursor is on a headline, but not on a link, offer all links in the headline and entry text. If you want to setup the frame configuration for following links, customize `org-link-frame-setup`.

RET When `org-return-follows-link` is set, `RET` will also follow the link at point.

mouse-2
mouse-1 On links, *mouse-1* and *mouse-2* will open the link just as *C-c C-o* would.

mouse-3 Like *mouse-2*, but force file links to be opened with Emacs, and internal links to be displayed in another window[10].

C-c C-x C-v `org-toggle-inline-images`
Toggle the inline display of linked images. Normally this will only inline images that have no description part in the link, i.e., images that will also be inlined during export. When called with a prefix argument, also display images that do have a link description. You can ask for inline images to be displayed at startup by configuring the variable `org-startup-with-inline-images`[11].

C-c % `org-mark-ring-push`
Push the current position onto the mark ring, to be able to return easily. Commands following an internal link do this automatically.

C-c & `org-mark-ring-goto`
Jump back to a recorded position. A position is recorded by the commands following internal links, and by *C-c %*. Using this command several times in direct succession moves through a ring of previously recorded positions.

C-c C-x C-n `org-next-link`
C-c C-x C-p `org-previous-link`
Move forward/backward to the next link in the buffer. At the limit of the buffer, the search fails once, and then wraps around. The key bindings for this are really too long; you might want to bind this also to *C-n* and *C-p*

```
(add-hook 'org-load-hook
  (lambda ()
    (define-key org-mode-map "\C-n" 'org-next-link)
    (define-key org-mode-map "\C-p" 'org-previous-link)))
```

[10] See the option `org-display-internal-link-with-indirect-buffer`

[11] with corresponding `#+STARTUP` keywords `inlineimages` and `noinlineimages`

4.5 Using links outside Org

You can insert and follow links that have Org syntax not only in Org, but in any Emacs buffer. For this, you should create two global commands, like this (please select suitable global keys yourself):

```
(global-set-key "\C-c L" 'org-insert-link-global)
(global-set-key "\C-c o" 'org-open-at-point-global)
```

4.6 Link abbreviations

Long URLs can be cumbersome to type, and often many similar links are needed in a document. For this you can use link abbreviations. An abbreviated link looks like this

```
[[linkword:tag][description]]
```

where the tag is optional. The *linkword* must be a word, starting with a letter, followed by letters, numbers, '-', and '_'. Abbreviations are resolved according to the information in the variable `org-link-abbrev-alist` that relates the linkwords to replacement text. Here is an example:

```
(setq org-link-abbrev-alist
  '(("bugzilla"  . "http://10.1.2.9/bugzilla/show_bug.cgi?id=")
    ("url-to-ja" . "http://translate.google.fr/translate?sl=en&tl=ja&u=%h")
    ("google"    . "http://www.google.com/search?q=")
    ("gmap"      . "http://maps.google.com/maps?q=%s")
    ("omap"      . "http://nominatim.openstreetmap.org/search?q=%s&polygon=1")
    ("ads"       . "http://adsabs.harvard.edu/cgi-bin/nph-abs_connect?author=%s&db_key=AST")))
```

If the replacement text contains the string '`%s`', it will be replaced with the tag. Using '`%h`' instead of '`%s`' will url-encode the tag (see the example above, where we need to encode the URL parameter.) Using '`%(my-function)`' will pass the tag to a custom function, and replace it by the resulting string.

If the replacement text doesn't contain any specifier, the tag will simply be appended in order to create the link.

Instead of a string, you may also specify a function that will be called with the tag as the only argument to create the link.

With the above setting, you could link to a specific bug with `[[bugzilla:129]]`, search the web for '`OrgMode`' with `[[google:OrgMode]]`, show the map location of the Free Software Foundation `[[gmap:51 Franklin Street, Boston]]` or of Carsten office `[[omap:Science Park 904, Amsterdam, The Netherlands]]` and find out what the Org author is doing besides Emacs hacking with `[[ads:Dominik,C]]`.

If you need special abbreviations just for a single Org buffer, you can define them in the file with

```
#+LINK: bugzilla  http://10.1.2.9/bugzilla/show_bug.cgi?id=
#+LINK: google    http://www.google.com/search?q=%s
```

In-buffer completion (see Section 15.1 [Completion], page 228) can be used after '`[`' to complete link abbreviations. You may also define a function that implements special (e.g., completion) support for inserting such a link with *C-c C-l*. Such a function should not accept any arguments, and return the full link with prefix. You can add a completion function to a link like this:

```
(org-link-set-parameters ``type'' :complete #'some-function)
```

4.7 Search options in file links

File links can contain additional information to make Emacs jump to a particular location in the file when following a link. This can be a line number or a search option after a double[12] colon. For example, when the command `C-c l` creates a link (see Section 4.4 [Handling links], page 41) to a file, it encodes the words in the current line as a search string that can be used to find this line back later when following the link with `C-c C-o`.

Here is the syntax of the different ways to attach a search to a file link, together with an explanation:

```
[[file:~/code/main.c::255]]
[[file:~/xx.org::My Target]]
[[file:~/xx.org::*My Target]]
[[file:~/xx.org::#my-custom-id]]
[[file:~/xx.org::/regexp/]]
```

`255` Jump to line 255.

`My Target` Search for a link target '<<My Target>>', or do a text search for '`my target`', similar to the search in internal links, see Section 4.2 [Internal links], page 38. In HTML export (see Section 12.9 [HTML export], page 149), such a file link will become an HTML reference to the corresponding named anchor in the linked file.

`*My Target`
In an Org file, restrict search to headlines.

`#my-custom-id`
Link to a heading with a `CUSTOM_ID` property

`/regexp/` Do a regular expression search for `regexp`. This uses the Emacs command `occur` to list all matches in a separate window. If the target file is in Org mode, `org-occur` is used to create a sparse tree with the matches.

As a degenerate case, a file link with an empty file name can be used to search the current file. For example, `[[file:::find me]]` does a search for '`find me`' in the current file, just as '`[[find me]]`' would.

4.8 Custom Searches

The default mechanism for creating search strings and for doing the actual search related to a file link may not work correctly in all cases. For example, BibTeX database files have many entries like '`year="1993"`' which would not result in good search strings, because the only unique identification for a BibTeX entry is the citation key.

If you come across such a problem, you can write custom functions to set the right search string for a particular file type, and to do the search for the string in the file. Using `add-hook`, these functions need to be added to the hook variables `org-create-file-search-functions` and `org-execute-file-search-functions`. See the docstring for these variables for more information. Org actually uses this mechanism for BibTeX database files, and you can use the corresponding code as an implementation example. See the file `org-bibtex.el`.

---

[12] For backward compatibility, line numbers can also follow a single colon.

5 TODO items

Org mode does not maintain TODO lists as separate documents[1]. Instead, TODO items are an integral part of the notes file, because TODO items usually come up while taking notes! With Org mode, simply mark any entry in a tree as being a TODO item. In this way, information is not duplicated, and the entire context from which the TODO item emerged is always present.

Of course, this technique for managing TODO items scatters them throughout your notes file. Org mode compensates for this by providing methods to give you an overview of all the things that you have to do.

5.1 Basic TODO functionality

Any headline becomes a TODO item when it starts with the word '`TODO`', for example:

```
*** TODO Write letter to Sam Fortune
```

The most important commands to work with TODO entries are:

C-c C-t `org-todo`

Rotate the TODO state of the current item among

```
,-> (unmarked) -> TODO -> DONE --.
'--------------------------------'
```

If TODO keywords have fast access keys (see Section 5.2.4 [Fast access to TODO states], page 49), you will be prompted for a TODO keyword through the fast selection interface; this is the default behavior when **`org-use-fast-todo-selection`** is non-**`nil`**.

The same rotation can also be done "remotely" from agenda buffers with the *t* command key (see Section 10.5 [Agenda commands], page 115).

C-u C-c C-t

When TODO keywords have no selection keys, select a specific keyword using completion; otherwise force cycling through TODO states with no prompt. When **`org-use-fast-todo-selection`** is set to **`prefix`**, use the fast selection interface.

S-right / S-left

Select the following/preceding TODO state, similar to cycling. Useful mostly if more than two TODO states are possible (see Section 5.2 [TODO extensions], page 47). See also Section 15.10.2 [Conflicts], page 238, for a discussion of the interaction with **`shift-selection-mode`**. See also the variable **`org-treat-S-cursor-todo-selection-as-state-change`**.

C-c / t `org-show-todo-tree`

View TODO items in a *sparse tree* (see Section 2.6 [Sparse trees], page 11). Folds the entire buffer, but shows all TODO items (with not-DONE state) and the headings hierarchy above them. With a prefix argument (or by using *C-c*

[1] Of course, you can make a document that contains only long lists of TODO items, but this is not required.

/ T), search for a specific TODO. You will be prompted for the keyword, and you can also give a list of keywords like `KWD1|KWD2|...` to list entries that match any one of these keywords. With a numeric prefix argument N, show the tree for the Nth keyword in the option `org-todo-keywords`. With two prefix arguments, find all TODO states, both un-done and done.

C-c a t `org-todo-list`
: Show the global TODO list. Collects the TODO items (with not-DONE states) from all agenda files (see Chapter 10 [Agenda views], page 102) into a single buffer. The new buffer will be in `agenda-mode`, which provides commands to examine and manipulate the TODO entries from the new buffer (see Section 10.5 [Agenda commands], page 115). See Section 10.3.2 [Global TODO list], page 106, for more information.

S-M-RET `org-insert-todo-heading`
: Insert a new TODO entry below the current one.

Changing a TODO state can also trigger tag changes. See the docstring of the option `org-todo-state-tags-triggers` for details.

5.2 Extended use of TODO keywords

By default, marked TODO entries have one of only two states: TODO and DONE. Org mode allows you to classify TODO items in more complex ways with *TODO keywords* (stored in `org-todo-keywords`). With special setup, the TODO keyword system can work differently in different files.

Note that *tags* are another way to classify headlines in general and TODO items in particular (see Chapter 6 [Tags], page 59).

5.2.1 TODO keywords as workflow states

You can use TODO keywords to indicate different *sequential* states in the process of working on an item, for example[2]:

```
(setq org-todo-keywords
  '((sequence "TODO" "FEEDBACK" "VERIFY" "|" "DONE" "DELEGATED")))
```

The vertical bar separates the TODO keywords (states that *need action*) from the DONE states (which need *no further action*). If you don't provide the separator bar, the last state is used as the DONE state. With this setup, the command *C-c C-t* will cycle an entry from TODO to FEEDBACK, then to VERIFY, and finally to DONE and DELEGATED. You may also use a numeric prefix argument to quickly select a specific state. For example *C-3 C-c C-t* will change the state immediately to VERIFY. Or you can use *S-left* to go backward through the sequence. If you define many keywords, you can use in-buffer completion (see Section 15.1 [Completion], page 228) or even a special one-key selection scheme (see Section 5.2.4 [Fast access to TODO states], page 49) to insert these words into the buffer. Changing a TODO state can be logged with a timestamp, see Section 5.3.2 [Tracking TODO state changes], page 52, for more information.

---

[2] Changing this variable only becomes effective after restarting Org mode in a buffer.

5.2.2 TODO keywords as types

The second possibility is to use TODO keywords to indicate different *types* of action items. For example, you might want to indicate that items are for "work" or "home". Or, when you work with several people on a single project, you might want to assign action items directly to persons, by using their names as TODO keywords. This would be set up like this:

```
(setq org-todo-keywords '((type "Fred" "Sara" "Lucy" "|" "DONE")))
```

In this case, different keywords do not indicate a sequence, but rather different types. So the normal work flow would be to assign a task to a person, and later to mark it DONE. Org mode supports this style by adapting the workings of the command `C-c C-t`[3]. When used several times in succession, it will still cycle through all names, in order to first select the right type for a task. But when you return to the item after some time and execute `C-c C-t` again, it will switch from any name directly to DONE. Use prefix arguments or completion to quickly select a specific name. You can also review the items of a specific TODO type in a sparse tree by using a numeric prefix to `C-c / t`. For example, to see all things Lucy has to do, you would use `C-3 C-c / t`. To collect Lucy's items from all agenda files into a single buffer, you would use the numeric prefix argument as well when creating the global TODO list: `C-3 C-c a t`.

5.2.3 Multiple keyword sets in one file

Sometimes you may want to use different sets of TODO keywords in parallel. For example, you may want to have the basic `TODO`/`DONE`, but also a workflow for bug fixing, and a separate state indicating that an item has been canceled (so it is not DONE, but also does not require action). Your setup would then look like this:

```
(setq org-todo-keywords
      '((sequence "TODO" "|" "DONE")
        (sequence "REPORT" "BUG" "KNOWNCAUSE" "|" "FIXED")
        (sequence "|" "CANCELED")))
```

The keywords should all be different, this helps Org mode to keep track of which subsequence should be used for a given entry. In this setup, `C-c C-t` only operates within a subsequence, so it switches from `DONE` to (nothing) to `TODO`, and from `FIXED` to (nothing) to `REPORT`. Therefore you need a mechanism to initially select the correct sequence. Besides the obvious ways like typing a keyword or using completion, you may also apply the following commands:

`C-u C-u C-c C-t`
`C-S-right`
`C-S-left` These keys jump from one TODO subset to the next. In the above example, `C-u C-u C-c C-t` or `C-S-right` would jump from `TODO` or `DONE` to `REPORT`, and any of the words in the second row to `CANCELED`. Note that the `C-S-` key binding conflict with `shift-selection-mode` (see Section 15.10.2 [Conflicts], page 238).

[3] This is also true for the `t` command in the agenda buffers.

S-right
S-left *S-left* and *S-right* and walk through *all* keywords from all sets, so for example *S-right* would switch from DONE to REPORT in the example above. See also Section 15.10.2 [Conflicts], page 238, for a discussion of the interaction with shift-selection-mode.

5.2.4 Fast access to TODO states

If you would like to quickly change an entry to an arbitrary TODO state instead of cycling through the states, you can set up keys for single-letter access to the states. This is done by adding the selection character after each keyword, in parentheses[4]. For example:

```
(setq org-todo-keywords
      '((sequence "TODO(t)" "|" "DONE(d)")
        (sequence "REPORT(r)" "BUG(b)" "KNOWNCAUSE(k)" "|" "FIXED(f)")
        (sequence "|" "CANCELED(c)")))
```

If you then press *C-c C-t* followed by the selection key, the entry will be switched to this state. *SPC* can be used to remove any TODO keyword from an entry.[5]

5.2.5 Setting up keywords for individual files

It can be very useful to use different aspects of the TODO mechanism in different files. For file-local settings, you need to add special lines to the file which set the keywords and interpretation for that file only. For example, to set one of the two examples discussed above, you need one of the following lines anywhere in the file:

```
#+TODO: TODO FEEDBACK VERIFY | DONE CANCELED
```

(you may also write #+SEQ_TODO to be explicit about the interpretation, but it means the same as #+TODO), or

```
#+TYP_TODO: Fred Sara Lucy Mike | DONE
```

A setup for using several sets in parallel would be:

```
#+TODO: TODO | DONE
#+TODO: REPORT BUG KNOWNCAUSE | FIXED
#+TODO: | CANCELED
```

To make sure you are using the correct keyword, type '#+' into the buffer and then use *M-TAB* completion.

Remember that the keywords after the vertical bar (or the last keyword if no bar is there) must always mean that the item is DONE (although you may use a different word). After changing one of these lines, use *C-c C-c* with the cursor still in the line to make the changes known to Org mode[6].

---

[4] All characters are allowed except @^!, which have a special meaning here.

[5] Check also the option org-fast-tag-selection-include-todo, it allows you to change the TODO state through the tags interface (see Section 6.2 [Setting tags], page 59), in case you like to mingle the two concepts. Note that this means you need to come up with unique keys across both sets of keywords.

[6] Org mode parses these lines only when Org mode is activated after visiting a file. *C-c C-c* with the cursor in a line starting with '#+' is simply restarting Org mode for the current buffer.

5.2.6 Faces for TODO keywords

Org mode highlights TODO keywords with special faces: `org-todo` for keywords indicating that an item still has to be acted upon, and `org-done` for keywords indicating that an item is finished. If you are using more than 2 different states, you might want to use special faces for some of them. This can be done using the option `org-todo-keyword-faces`. For example:

```
(setq org-todo-keyword-faces
      '(("TODO" . org-warning) ("STARTED" . "yellow")
        ("CANCELED" . (:foreground "blue" :weight bold))))
```

While using a list with face properties as shown for CANCELED *should* work, this does not always seem to be the case. If necessary, define a special face and use that. A string is interpreted as a color. The option `org-faces-easy-properties` determines if that color is interpreted as a foreground or a background color.

5.2.7 TODO dependencies

The structure of Org files (hierarchy and lists) makes it easy to define TODO dependencies. Usually, a parent TODO task should not be marked DONE until all subtasks (defined as children tasks) are marked as DONE. And sometimes there is a logical sequence to a number of (sub)tasks, so that one task cannot be acted upon before all siblings above it are done. If you customize the option `org-enforce-todo-dependencies`, Org will block entries from changing state to DONE while they have children that are not DONE. Furthermore, if an entry has a property `ORDERED`, each of its children will be blocked until all earlier siblings are marked DONE. Here is an example:

```
* TODO Blocked until (two) is done
** DONE one
** TODO two

* Parent
  :PROPERTIES:
  :ORDERED: t
  :END:
** TODO a
** TODO b, needs to wait for (a)
** TODO c, needs to wait for (a) and (b)
```

You can ensure an entry is never blocked by using the `NOBLOCKING` property:

```
* This entry is never blocked
  :PROPERTIES:
  :NOBLOCKING: t
  :END:
```

C-c C-x o — `org-toggle-ordered-property`

Toggle the `ORDERED` property of the current entry. A property is used for this behavior because this should be local to the current entry, not inherited like a tag. However, if you would like to *track* the value of this property with a tag for better visibility, customize the option `org-track-ordered-property-with-tag`.

C-u C-u C-u C-c C-t
Change TODO state, circumventing any state blocking.

If you set the option `org-agenda-dim-blocked-tasks`, TODO entries that cannot be closed because of such dependencies will be shown in a dimmed font or even made invisible in agenda views (see Chapter 10 [Agenda views], page 102).

You can also block changes of TODO states by looking at checkboxes (see Section 5.6 [Checkboxes], page 56). If you set the option `org-enforce-todo-checkbox-dependencies`, an entry that has unchecked checkboxes will be blocked from switching to DONE.

If you need more complex dependency structures, for example dependencies between entries in different trees or files, check out the contributed module `org-depend.el`.

5.3 Progress logging

Org mode can automatically record a timestamp and possibly a note when you mark a TODO item as DONE, or even each time you change the state of a TODO item. This system is highly configurable; settings can be on a per-keyword basis and can be localized to a file or even a subtree. For information on how to clock working time for a task, see Section 8.4 [Clocking work time], page 80.

5.3.1 Closing items

The most basic logging is to keep track of *when* a certain TODO item was finished. This is achieved with[1]

```
(setq org-log-done 'time)
```

Then each time you turn an entry from a TODO (not-done) state into any of the DONE states, a line '`CLOSED: [timestamp]`' will be inserted just after the headline. If you turn the entry back into a TODO item through further state cycling, that line will be removed again. If you turn the entry back to a non-TODO state (by pressing `C-c C-t SPC` for example), that line will also be removed, unless you set `org-closed-keep-when-no-todo` to non-`nil`. If you want to record a note along with the timestamp, use[2]

```
(setq org-log-done 'note)
```

You will then be prompted for a note, and that note will be stored below the entry with a '`Closing Note`' heading.

5.3.2 Tracking TODO state changes

When TODO keywords are used as workflow states (see Section 5.2.1 [Workflow states], page 47), you might want to keep track of when a state change occurred and maybe take a note about this change. You can either record just a timestamp, or a time-stamped note for a change. These records will be inserted after the headline as an itemized list, newest first[3]. When taking a lot of notes, you might want to get the notes out of the way into a drawer (see Section 2.8 [Drawers], page 15). Customize `org-log-into-drawer` to get this behavior—the recommended drawer for this is called `LOGBOOK`[4]. You can also overrule the setting of this variable for a subtree by setting a `LOG_INTO_DRAWER` property.

Since it is normally too much to record a note for every state, Org mode expects configuration on a per-keyword basis for this. This is achieved by adding special markers '`!`' (for a timestamp) or '`@`' (for a note with timestamp) in parentheses after each keyword. For example, with the setting

```
(setq org-todo-keywords
  '((sequence "TODO(t)" "WAIT(w@/!)" "|" "DONE(d!)" "CANCELED(c@)")))
```

To record a timestamp without a note for TODO keywords configured with '`@`', just type *C-c C-c* to enter a blank note when prompted.

---

[1] The corresponding in-buffer setting is: `#+STARTUP: logdone`

[2] The corresponding in-buffer setting is: `#+STARTUP: lognotedone`.

[3] See the option `org-log-states-order-reversed`

[4] Note that the `LOGBOOK` drawer is unfolded when pressing `SPC` in the agenda to show an entry—use `C-u SPC` to keep it folded here

You not only define global TODO keywords and fast access keys, but also request that a time is recorded when the entry is set to DONE[5], and that a note is recorded when switching to WAIT or CANCELED. The setting for WAIT is even more special: the '`!`' after the slash means that in addition to the note taken when entering the state, a timestamp should be recorded when *leaving* the WAIT state, if and only if the *target* state does not configure logging for entering it. So it has no effect when switching from WAIT to DONE, because DONE is configured to record a timestamp only. But when switching from WAIT back to TODO, the '`/!`' in the WAIT setting now triggers a timestamp even though TODO has no logging configured.

You can use the exact same syntax for setting logging preferences local to a buffer:

```
#+TODO: TODO(t) WAIT(w@/!) | DONE(d!) CANCELED(c@)
```

In order to define logging settings that are local to a subtree or a single item, define a LOGGING property in this entry. Any non-empty LOGGING property resets all logging settings to `nil`. You may then turn on logging for this specific tree using STARTUP keywords like `lognotedone` or `logrepeat`, as well as adding state specific settings like `TODO(!)`. For example

```
* TODO Log each state with only a time
  :PROPERTIES:
  :LOGGING: TODO(!) WAIT(!) DONE(!) CANCELED(!)
  :END:
* TODO Only log when switching to WAIT, and when repeating
  :PROPERTIES:
  :LOGGING: WAIT(@) logrepeat
  :END:
* TODO No logging at all
  :PROPERTIES:
  :LOGGING: nil
  :END:
```

5.3.3 Tracking your habits

Org has the ability to track the consistency of a special category of TODOs, called "habits". A habit has the following properties:

1. You have enabled the `habits` module by customizing `org-modules`.
2. The habit is a TODO item, with a TODO keyword representing an open state.
3. The property `STYLE` is set to the value `habit`.
4. The TODO has a scheduled date, usually with a `.+` style repeat interval. A `++` style may be appropriate for habits with time constraints, e.g., must be done on weekends, or a `+` style for an unusual habit that can have a backlog, e.g., weekly reports.
5. The TODO may also have minimum and maximum ranges specified by using the syntax '`.+2d/3d`', which says that you want to do the task at least every three days, but at most every two days.

---

[5] It is possible that Org mode will record two timestamps when you are using both `org-log-done` and state change logging. However, it will never prompt for two notes—if you have configured both, the state change recording note will take precedence and cancel the '`Closing Note`'.

6. You must also have state logging for the `DONE` state enabled (see Section 5.3.2 [Tracking TODO state changes], page 52), in order for historical data to be represented in the consistency graph. If it is not enabled it is not an error, but the consistency graphs will be largely meaningless.

To give you an idea of what the above rules look like in action, here's an actual habit with some history:

```
** TODO Shave
   SCHEDULED: <2009-10-17 Sat .+2d/4d>
   :PROPERTIES:
   :STYLE:    habit
   :LAST_REPEAT: [2009-10-19 Mon 00:36]
   :END:
   - State "DONE"       from "TODO"       [2009-10-15 Thu]
   - State "DONE"       from "TODO"       [2009-10-12 Mon]
   - State "DONE"       from "TODO"       [2009-10-10 Sat]
   - State "DONE"       from "TODO"       [2009-10-04 Sun]
   - State "DONE"       from "TODO"       [2009-10-02 Fri]
   - State "DONE"       from "TODO"       [2009-09-29 Tue]
   - State "DONE"       from "TODO"       [2009-09-25 Fri]
   - State "DONE"       from "TODO"       [2009-09-19 Sat]
   - State "DONE"       from "TODO"       [2009-09-16 Wed]
   - State "DONE"       from "TODO"       [2009-09-12 Sat]
```

What this habit says is: I want to shave at most every 2 days (given by the `SCHEDULED` date and repeat interval) and at least every 4 days. If today is the 15th, then the habit first appears in the agenda on Oct 17, after the minimum of 2 days has elapsed, and will appear overdue on Oct 19, after four days have elapsed.

What's really useful about habits is that they are displayed along with a consistency graph, to show how consistent you've been at getting that task done in the past. This graph shows every day that the task was done over the past three weeks, with colors for each day. The colors used are:

`Blue` If the task wasn't to be done yet on that day.

`Green` If the task could have been done on that day.

`Yellow` If the task was going to be overdue the next day.

`Red` If the task was overdue on that day.

In addition to coloring each day, the day is also marked with an asterisk if the task was actually done that day, and an exclamation mark to show where the current day falls in the graph.

There are several configuration variables that can be used to change the way habits are displayed in the agenda.

`org-habit-graph-column`
: The buffer column at which the consistency graph should be drawn. This will overwrite any text in that column, so it is a good idea to keep your habits' titles brief and to the point.

`org-habit-preceding-days`
: The amount of history, in days before today, to appear in consistency graphs.

`org-habit-following-days`
: The number of days after today that will appear in consistency graphs.

`org-habit-show-habits-only-for-today`
: If non-`nil`, only show habits in today's agenda view. This is set to true by default.

Lastly, pressing *K* in the agenda buffer will cause habits to temporarily be disabled and they won't appear at all. Press *K* again to bring them back. They are also subject to tag filtering, if you have habits which should only be done in certain contexts, for example.

5.4 Priorities

If you use Org mode extensively, you may end up with enough TODO items that it starts to make sense to prioritize them. Prioritizing can be done by placing a *priority cookie* into the headline of a TODO item, like this

```
*** TODO [#A] Write letter to Sam Fortune
```

By default, Org mode supports three priorities: '`A`', '`B`', and '`C`'. '`A`' is the highest priority. An entry without a cookie is treated just like priority '`B`'. Priorities make a difference only for sorting in the agenda (see Section 10.3.1 [Weekly/daily agenda], page 104); outside the agenda, they have no inherent meaning to Org mode. The cookies can be highlighted with special faces by customizing `org-priority-faces`.

Priorities can be attached to any outline node; they do not need to be TODO items.

C-c ,
: Set the priority of the current headline (`org-priority`). The command prompts for a priority character '`A`', '`B`' or '`C`'. When you press `SPC` instead, the priority cookie is removed from the headline. The priorities can also be changed "remotely" from the agenda buffer with the `,` command (see Section 10.5 [Agenda commands], page 115).

S-up — `org-priority-up`
S-down — `org-priority-down`
: Increase/decrease priority of current headline[6]. Note that these keys are also used to modify timestamps (see Section 8.2 [Creating timestamps], page 74). See also Section 15.10.2 [Conflicts], page 238, for a discussion of the interaction with `shift-selection-mode`.

You can change the range of allowed priorities by setting the options `org-highest-priority`, `org-lowest-priority`, and `org-default-priority`. For an individual buffer, you may set these values (highest, lowest, default) like this (please make sure that the highest priority is earlier in the alphabet than the lowest priority):

```
#+PRIORITIES: A C B
```

[6] See also the option `org-priority-start-cycle-with-default`.

5.5 Breaking tasks down into subtasks

It is often advisable to break down large tasks into smaller, manageable subtasks. You can do this by creating an outline tree below a TODO item, with detailed subtasks on the tree[7]. To keep the overview over the fraction of subtasks that are already completed, insert either '`[/]`' or '`[%]`' anywhere in the headline. These cookies will be updated each time the TODO status of a child changes, or when pressing *C-c C-c* on the cookie. For example:

```
* Organize Party [33%]
** TODO Call people [1/2]
*** TODO Peter
*** DONE Sarah
** TODO Buy food
** DONE Talk to neighbor
```

If a heading has both checkboxes and TODO children below it, the meaning of the statistics cookie become ambiguous. Set the property `COOKIE_DATA` to either '`checkbox`' or '`todo`' to resolve this issue.

If you would like to have the statistics cookie count any TODO entries in the subtree (not just direct children), configure `org-hierarchical-todo-statistics`. To do this for a single subtree, include the word '`recursive`' into the value of the `COOKIE_DATA` property.

```
* Parent capturing statistics [2/20]
  :PROPERTIES:
  :COOKIE_DATA: todo recursive
  :END:
```

If you would like a TODO entry to automatically change to DONE when all children are done, you can use the following setup:

```
(defun org-summary-todo (n-done n-not-done)
  "Switch entry to DONE when all subentries are done, to TODO otherwise."
  (let (org-log-done org-log-states)   ; turn off logging
    (org-todo (if (= n-not-done 0) "DONE" "TODO"))))

(add-hook 'org-after-todo-statistics-hook 'org-summary-todo)
```

Another possibility is the use of checkboxes to identify (a hierarchy of) a large number of subtasks (see Section 5.6 [Checkboxes], page 56).

5.6 Checkboxes

Every item in a plain list[8] (see Section 2.7 [Plain lists], page 12) can be made into a checkbox by starting it with the string '`[ ]`'. This feature is similar to TODO items (see Chapter 5 [TODO items], page 46), but is more lightweight. Checkboxes are not included in the global TODO list, so they are often great to split a task into a number of simple steps. Or you can use them in a shopping list. To toggle a checkbox, use *C-c C-c*, or use the mouse (thanks to Piotr Zielinski's `org-mouse.el`).

Here is an example of a checkbox list.

---

[7] To keep subtasks out of the global TODO list, see the `org-agenda-todo-list-sublevels`.

[8] With the exception of description lists. But you can allow it by modifying `org-list-automatic-rules` accordingly.

```
* TODO Organize party [2/4]
  - [-] call people [1/3]
    - [ ] Peter
    - [X] Sarah
    - [ ] Sam
  - [X] order food
  - [ ] think about what music to play
  - [X] talk to the neighbors
```

Checkboxes work hierarchically, so if a checkbox item has children that are checkboxes, toggling one of the children checkboxes will make the parent checkbox reflect if none, some, or all of the children are checked.

The '[2/4]' and '[1/3]' in the first and second line are cookies indicating how many checkboxes present in this entry have been checked off, and the total number of checkboxes present. This can give you an idea on how many checkboxes remain, even without opening a folded entry. The cookies can be placed into a headline or into (the first line of) a plain list item. Each cookie covers checkboxes of direct children structurally below the headline/item on which the cookie appears[9]. You have to insert the cookie yourself by typing either '[/]' or '[%]'. With '[/]' you get an 'n out of m' result, as in the examples above. With '[%]' you get information about the percentage of checkboxes checked (in the above example, this would be '[50%]' and '[33%]', respectively). In a headline, a cookie can count either checkboxes below the heading or TODO states of children, and it will display whatever was changed last. Set the property `COOKIE_DATA` to either 'checkbox' or 'todo' to resolve this issue.

If the current outline node has an `ORDERED` property, checkboxes must be checked off in sequence, and an error will be thrown if you try to check off a box while there are unchecked boxes above it.

The following commands work with checkboxes:

C-c C-c `org-toggle-checkbox`
: Toggle checkbox status or (with prefix arg) checkbox presence at point. With a single prefix argument, add an empty checkbox or remove the current one[10]. With a double prefix argument, set it to '[-]', which is considered to be an intermediate state.

C-c C-x C-b `org-toggle-checkbox`
: Toggle checkbox status or (with prefix arg) checkbox presence at point. With double prefix argument, set it to '[-]', which is considered to be an intermediate state.

  - If there is an active region, toggle the first checkbox in the region and set all remaining boxes to the same status as the first. With a prefix arg, add or remove the checkbox for all items in the region.
  - If the cursor is in a headline, toggle the state of the first checkbox in the region between this headline and the next—so *not* the entire subtree—and propagate this new state to all other checkboxes in the same area.

---

[9] Set the option `org-checkbox-hierarchical-statistics` if you want such cookies to count all checkboxes below the cookie, not just those belonging to direct children.

[10] *C-u C-c C-c* before the *first* bullet in a list with no checkbox will add checkboxes to the rest of the list.

– If there is no active region, just toggle the checkbox at point.

M-S-RET `org-insert-todo-heading`
Insert a new item with a checkbox. This works only if the cursor is already in a plain list item (see Section 2.7 [Plain lists], page 12).

C-c C-x o `org-toggle-ordered-property`
Toggle the `ORDERED` property of the entry, to toggle if checkboxes must be checked off in sequence. A property is used for this behavior because this should be local to the current entry, not inherited like a tag. However, if you would like to *track* the value of this property with a tag for better visibility, customize `org-track-ordered-property-with-tag`.

C-c # `org-update-statistics-cookies`
Update the statistics cookie in the current outline entry. When called with a *C-u* prefix, update the entire file. Checkbox statistic cookies are updated automatically if you toggle checkboxes with *C-c C-c* and make new ones with *M-S-RET*. TODO statistics cookies update when changing TODO states. If you delete boxes/entries or add/change them by hand, use this command to get things back into sync.

6 Tags

An excellent way to implement labels and contexts for cross-correlating information is to assign *tags* to headlines. Org mode has extensive support for tags.

Every headline can contain a list of tags; they occur at the end of the headline. Tags are normal words containing letters, numbers, '_', and '@'. Tags must be preceded and followed by a single colon, e.g., ':work:'. Several tags can be specified, as in ':work:urgent:'. Tags will by default be in bold face with the same color as the headline. You may specify special faces for specific tags using the option `org-tag-faces`, in much the same way as you can for TODO keywords (see Section 5.2.6 [Faces for TODO keywords], page 50).

6.1 Tag inheritance

Tags make use of the hierarchical structure of outline trees. If a heading has a certain tag, all subheadings will inherit the tag as well. For example, in the list

```
* Meeting with the French group      :work:
** Summary by Frank                  :boss:notes:
*** TODO Prepare slides for him      :action:
```

the final heading will have the tags ':work:', ':boss:', ':notes:', and ':action:' even though the final heading is not explicitly marked with all those tags. You can also set tags that all entries in a file should inherit just as if these tags were defined in a hypothetical level zero that surrounds the entire file. Use a line like this[1]:

```
#+FILETAGS: :Peter:Boss:Secret:
```

To limit tag inheritance to specific tags, use `org-tags-exclude-from-inheritance`. To turn it off entirely, use `org-use-tag-inheritance`.

When a headline matches during a tags search while tag inheritance is turned on, all the sublevels in the same tree will (for a simple match form) match as well[2]. The list of matches may then become very long. If you only want to see the first tags match in a subtree, configure `org-tags-match-list-sublevels` (not recommended).

Tag inheritance is relevant when the agenda search tries to match a tag, either in the `tags` or `tags-todo` agenda types. In other agenda types, `org-use-tag-inheritance` has no effect. Still, you may want to have your tags correctly set in the agenda, so that tag filtering works fine, with inherited tags. Set `org-agenda-use-tag-inheritance` to control this: the default value includes all agenda types, but setting this to `nil` can really speed up agenda generation.

6.2 Setting tags

Tags can simply be typed into the buffer at the end of a headline. After a colon, *M-TAB* offers completion on tags. There is also a special command for inserting tags:

C-c C-q `org-set-tags-command`

Enter new tags for the current headline. Org mode will either offer completion or a special single-key interface for setting tags, see below. After pressing `RET`,

---

[1] As with all these in-buffer settings, pressing *C-c C-c* activates any changes in the line.

[2] This is only true if the search does not involve more complex tests including properties (see Section 7.3 [Property searches], page 66).

the tags will be inserted and aligned to `org-tags-column`. When called with a *C-u* prefix, all tags in the current buffer will be aligned to that column, just to make things look nice. TAGS are automatically realigned after promotion, demotion, and TODO state changes (see Section 5.1 [TODO basics], page 46).

C-c C-c `org-set-tags-command`
When the cursor is in a headline, this does the same as *C-c C-q*.

Org supports tag insertion based on a *list of tags*. By default this list is constructed dynamically, containing all tags currently used in the buffer. You may also globally specify a hard list of tags with the variable `org-tag-alist`. Finally you can set the default tags for a given file with lines like

```
#+TAGS: @work @home @tennisclub
#+TAGS: laptop car pc sailboat
```

If you have globally defined your preferred set of tags using the variable `org-tag-alist`, but would like to use a dynamic tag list in a specific file, add an empty TAGS option line to that file:

```
#+TAGS:
```

If you have a preferred set of tags that you would like to use in every file, in addition to those defined on a per-file basis by TAGS option lines, then you may specify a list of tags with the variable `org-tag-persistent-alist`. You may turn this off on a per-file basis by adding a STARTUP option line to that file:

```
#+STARTUP: noptag
```

By default Org mode uses the standard minibuffer completion facilities for entering tags. However, it also implements another, quicker, tag selection method called *fast tag selection*. This allows you to select and deselect tags with just a single key press. For this to work well you should assign unique, case-sensitive, letters to most of your commonly used tags. You can do this globally by configuring the variable `org-tag-alist` in your Emacs init file. For example, you may find the need to tag many items in different files with '`:@home:`'. In this case you can set something like:

```
(setq org-tag-alist '(("@work" . ?w) ("@home" . ?h) ("laptop" . ?l)))
```

If the tag is only relevant to the file you are working on, then you can instead set the TAGS option line as:

```
#+TAGS: @work(w)  @home(h)  @tennisclub(t)  laptop(l)  pc(p)
```

The tags interface will show the available tags in a splash window. If you want to start a new line after a specific tag, insert '`\n`' into the tag list

```
#+TAGS: @work(w)  @home(h)  @tennisclub(t) \n laptop(l)  pc(p)
```

or write them in two lines:

```
#+TAGS: @work(w)  @home(h)  @tennisclub(t)
#+TAGS: laptop(l)  pc(p)
```

You can also group together tags that are mutually exclusive by using braces, as in:

```
#+TAGS: { @work(w)  @home(h)  @tennisclub(t) }  laptop(l)  pc(p)
```

you indicate that at most one of '`@work`', '`@home`', and '`@tennisclub`' should be selected. Multiple such groups are allowed.

Don't forget to press *C-c C-c* with the cursor in one of these lines to activate any changes.

To set these mutually exclusive groups in the variable `org-tag-alist`, you must use the dummy tags `:startgroup` and `:endgroup` instead of the braces. Similarly, you can use `:newline` to indicate a line break. The previous example would be set globally by the following configuration:

```
(setq org-tag-alist '((:startgroup . nil)
                      ("@work" . ?w) ("@home" . ?h)
                      ("@tennisclub" . ?t)
                      (:endgroup . nil)
                      ("laptop" . ?l) ("pc" . ?p)))
```

If at least one tag has a selection key then pressing *C-c C-c* will automatically present you with a special interface, listing inherited tags, the tags of the current headline, and a list of all valid tags with corresponding keys[3].

Pressing keys assigned to tags will add or remove them from the list of tags in the current line. Selecting a tag in a group of mutually exclusive tags will turn off any other tags from that group.

In this interface, you can also use the following special keys:

`TAB` Enter a tag in the minibuffer, even if the tag is not in the predefined list. You will be able to complete on all tags present in the buffer. You can also add several tags: just separate them with a comma.

`SPC` Clear all tags for this line.

`RET` Accept the modified set.

C-g Abort without installing changes.

q If *q* is not assigned to a tag, it aborts like *C-g*.

! Turn off groups of mutually exclusive tags. Use this to (as an exception) assign several tags from such a group.

C-c Toggle auto-exit after the next change (see below). If you are using expert mode, the first *C-c* will display the selection window.

This method lets you assign tags to a headline with very few keys. With the above setup, you could clear the current tags and set '`@home`', '`laptop`' and '`pc`' tags with just the following keys: *C-c C-c SPC h l p RET*. Switching from '`@home`' to '`@work`' would be done with *C-c C-c w RET* or alternatively with *C-c C-c C-c w*. Adding the non-predefined tag '`Sarah`' could be done with *C-c C-c TAB S a r a h RET RET*.

If you find that most of the time you need only a single key press to modify your list of tags, set `org-fast-tag-selection-single-key`. Then you no longer have to press `RET` to exit fast tag selection—it will immediately exit after the first change. If you then occasionally need more keys, press *C-c* to turn off auto-exit for the current tag selection process (in effect: start selection with *C-c C-c C-c* instead of *C-c C-c*). If you set the variable to the value `expert`, the special window is not even shown for single-key tag selection, it comes up only when you press an extra *C-c*.

---

[3] Keys will automatically be assigned to tags which have no configured keys.

6.3 Tag hierarchy

Tags can be defined in hierarchies. A tag can be defined as a *group tag* for a set of other tags. The group tag can be seen as the "broader term" for its set of tags. Defining multiple *group tags* and nesting them creates a tag hierarchy.

One use-case is to create a taxonomy of terms (tags) that can be used to classify nodes in a document or set of documents.

When you search for a group tag, it will return matches for all members in the group and its subgroups. In an agenda view, filtering by a group tag will display or hide headlines tagged with at least one of the members of the group or any of its subgroups. This makes tag searches and filters even more flexible.

You can set group tags by using brackets and inserting a colon between the group tag and its related tags—beware that all whitespaces are mandatory so that Org can parse this line correctly:

```
#+TAGS: [ GTD : Control Persp ]
```

In this example, 'GTD' is the *group tag* and it is related to two other tags: 'Control', 'Persp'. Defining 'Control' and 'Persp' as group tags creates an hierarchy of tags:

```
#+TAGS: [ Control : Context Task ]
#+TAGS: [ Persp : Vision Goal AOF Project ]
```

That can conceptually be seen as a hierarchy of tags:

```
- GTD
  - Persp
    - Vision
    - Goal
    - AOF
    - Project
  - Control
    - Context
    - Task
```

You can use the `:startgrouptag`, `:grouptags` and `:endgrouptag` keyword directly when setting `org-tag-alist` directly:

```
(setq org-tag-alist '((:startgrouptag)
                      ("GTD")
                      (:grouptags)
                      ("Control")
                      ("Persp")
                      (:endgrouptag)
                      (:startgrouptag)
                      ("Control")
                      (:grouptags)
                      ("Context")
                      ("Task")
                      (:endgrouptag)))
```

The tags in a group can be mutually exclusive if using the same group syntax as is used for grouping mutually exclusive tags together; using curly brackets.

```
#+TAGS: { Context : @Home @Work @Call }
```

When setting `org-tag-alist` you can use `:startgroup` & `:endgroup` instead of `:startgrouptag` & `:endgrouptag` to make the tags mutually exclusive.

Furthermore, the members of a *group tag* can also be regular expressions, creating the possibility of a more dynamic and rule-based tag structure. The regular expressions in the group must be specified within { }. Here is an expanded example:

```
#+TAGS: [ Vision : {V@.+} ]
#+TAGS: [ Goal : {G@.+} ]
#+TAGS: [ AOF : {AOF@.+} ]
#+TAGS: [ Project : {P@.+} ]
```

Searching for the tag '`Project`' will now list all tags also including regular expression matches for '`P@.+`', and similarly for tag searches on '`Vision`', '`Goal`' and '`AOF`'. For example, this would work well for a project tagged with a common project-identifier, e.g. '`P@2014_OrgTags`'.

If you want to ignore group tags temporarily, toggle group tags support with `org-toggle-tags-groups`, bound to *C-c C-x q*. If you want to disable tag groups completely, set `org-group-tags` to `nil`.

6.4 Tag searches

Once a system of tags has been set up, it can be used to collect related information into special lists.

C-c / m or *C-c * `org-match-sparse-tree`
: Create a sparse tree with all headlines matching a tags/property/TODO search. With a *C-u* prefix argument, ignore headlines that are not a TODO line. See Section 10.3.3 [Matching tags and properties], page 107.

C-c a m `org-tags-view`
: Create a global list of tag matches from all agenda files. See Section 10.3.3 [Matching tags and properties], page 107.

C-c a M `org-tags-view`
: Create a global list of tag matches from all agenda files, but check only TODO items and force checking subitems (see the option `org-tags-match-list-sublevels`).

These commands all prompt for a match string which allows basic Boolean logic like '`+boss+urgent-project1`', to find entries with tags '`boss`' and '`urgent`', but not '`project1`', or '`Kathy|Sally`' to find entries tagged as '`Kathy`' or '`Sally`'. The full syntax of the search string is rich and allows also matching against TODO keywords, entry levels and properties. For a complete description with many examples, see Section 10.3.3 [Matching tags and properties], page 107.

7 Properties and columns

A property is a key-value pair associated with an entry. Properties can be set so they are associated with a single entry, with every entry in a tree, or with every entry in an Org mode file.

There are two main applications for properties in Org mode. First, properties are like tags, but with a value. Imagine maintaining a file where you document bugs and plan releases for a piece of software. Instead of using tags like `:release_1:`, `:release_2:`, you can use a property, say `:Release:`, that in different subtrees has different values, such as `1.0` or `2.0`. Second, you can use properties to implement (very basic) database capabilities in an Org buffer. Imagine keeping track of your music CDs, where properties could be things such as the album, artist, date of release, number of tracks, and so on.

Properties can be conveniently edited and viewed in column view (see Section 7.5 [Column view], page 67).

7.1 Property syntax

Properties are key-value pairs. When they are associated with a single entry or with a tree they need to be inserted into a special drawer (see Section 2.8 [Drawers], page 15) with the name `PROPERTIES`, which has to be located right below a headline, and its planning line (see Section 8.3 [Deadlines and scheduling], page 77) when applicable. Each property is specified on a single line, with the key (surrounded by colons) first, and the value after it. Keys are case-insensitive. Here is an example:

```
* CD collection
** Classic
*** Goldberg Variations
    :PROPERTIES:
    :Title:     Goldberg Variations
    :Composer:  J.S. Bach
    :Artist:    Glen Gould
    :Publisher: Deutsche Grammophon
    :NDisks:    1
    :END:
```

Depending on the value of `org-use-property-inheritance`, a property set this way will either be associated with a single entry, or the subtree defined by the entry, see Section 7.4 [Property inheritance], page 67.

You may define the allowed values for a particular property '`:Xyz:`' by setting a property '`:Xyz_ALL:`'. This special property is *inherited*, so if you set it in a level 1 entry, it will apply to the entire tree. When allowed values are defined, setting the corresponding property becomes easier and is less prone to typing errors. For the example with the CD collection, we can predefine publishers and the number of disks in a box like this:

```
* CD collection
  :PROPERTIES:
  :NDisks_ALL:  1 2 3 4
  :Publisher_ALL: "Deutsche Grammophon" Philips EMI
  :END:
```

If you want to set properties that can be inherited by any entry in a file, use a line like

```
#+PROPERTY: NDisks_ALL 1 2 3 4
```

Contrary to properties set from a special drawer, you have to refresh the buffer with *C-c C-c* to activate this change.

If you want to add to the value of an existing property, append a `+` to the property name. The following results in the property `var` having the value "foo=1 bar=2".

```
#+PROPERTY: var  foo=1
#+PROPERTY: var+ bar=2
```

It is also possible to add to the values of inherited properties. The following results in the `genres` property having the value "Classic Baroque" under the `Goldberg Variations` subtree.

```
* CD collection
** Classic
    :PROPERTIES:
    :GENRES: Classic
    :END:
*** Goldberg Variations
    :PROPERTIES:
    :Title:     Goldberg Variations
    :Composer:  J.S. Bach
    :Artist:    Glen Gould
    :Publisher: Deutsche Grammophon
    :NDisks:    1
    :GENRES+:   Baroque
    :END:
```

Note that a property can only have one entry per Drawer.

Property values set with the global variable `org-global-properties` can be inherited by all entries in all Org files.

The following commands help to work with properties:

M-TAB `pcomplete`
: After an initial colon in a line, complete property keys. All keys used in the current file will be offered as possible completions.

C-c C-x p `org-set-property`
: Set a property. This prompts for a property name and a value. If necessary, the property drawer is created as well.

C-u M-x org-insert-drawer RET
: Insert a property drawer into the current entry. The drawer will be inserted early in the entry, but after the lines with planning information like deadlines.

C-c C-c `org-property-action`
: With the cursor in a property drawer, this executes property commands.

C-c C-c s `org-set-property`
: Set a property in the current entry. Both the property and the value can be inserted using completion.

S-right `org-property-next-allowed-value`
S-left `org-property-previous-allowed-value`
Switch property at point to the next/previous allowed value.

C-c C-c d `org-delete-property`
Remove a property from the current entry.

C-c C-c D `org-delete-property-globally`
Globally remove a property, from all entries in the current file.

C-c C-c c `org-compute-property-at-point`
Compute the property at point, using the operator and scope from the nearest column format definition.

7.2 Special properties

Special properties provide an alternative access method to Org mode features, like the TODO state or the priority of an entry, discussed in the previous chapters. This interface exists so that you can include these states in a column view (see Section 7.5 [Column view], page 67), or to use them in queries. The following property names are special and should not be used as keys in the properties drawer:

ALLTAGS	All tags, including inherited ones.
BLOCKED	"t" if task is currently blocked by children or siblings.
CLOCKSUM	The sum of CLOCK intervals in the subtree. `org-clock-sum` must be run first to compute the values in the current buffer.
CLOCKSUM_T	The sum of CLOCK intervals in the subtree for today. `org-clock-sum-today` must be run first to compute the values in the current buffer.
CLOSED	When was this entry closed?
DEADLINE	The deadline time string, without the angular brackets.
FILE	The filename the entry is located in.
ITEM	The headline of the entry.
PRIORITY	The priority of the entry, a string with a single letter.
SCHEDULED	The scheduling timestamp, without the angular brackets.
TAGS	The tags defined directly in the headline.
TIMESTAMP	The first keyword-less timestamp in the entry.
TIMESTAMP_IA	The first inactive timestamp in the entry.
TODO	The TODO keyword of the entry.

7.3 Property searches

To create sparse trees and special lists with selection based on properties, the same commands are used as for tag searches (see Section 6.4 [Tag searches], page 63).

C-c / m or *C-c * `org-match-sparse-tree`
Create a sparse tree with all matching entries. With a *C-u* prefix argument, ignore headlines that are not a TODO line.

C-c a m `org-tags-view`
Create a global list of tag/property matches from all agenda files. See Section 10.3.3 [Matching tags and properties], page 107.

`C-c a M` `org-tags-view`
: Create a global list of tag matches from all agenda files, but check only TODO items and force checking of subitems (see the option `org-tags-match-list-sublevels`).

The syntax for the search string is described in Section 10.3.3 [Matching tags and properties], page 107.

There is also a special command for creating sparse trees based on a single property:

`C-c / p`
: Create a sparse tree based on the value of a property. This first prompts for the name of a property, and then for a value. A sparse tree is created with all entries that define this property with the given value. If you enclose the value in curly braces, it is interpreted as a regular expression and matched against the property values.

7.4 Property Inheritance

The outline structure of Org mode documents lends itself to an inheritance model of properties: if the parent in a tree has a certain property, the children can inherit this property. Org mode does not turn this on by default, because it can slow down property searches significantly and is often not needed. However, if you find inheritance useful, you can turn it on by setting the variable `org-use-property-inheritance`. It may be set to `t` to make all properties inherited from the parent, to a list of properties that should be inherited, or to a regular expression that matches inherited properties. If a property has the value `nil`, this is interpreted as an explicit undefine of the property, so that inheritance search will stop at this value and return `nil`.

Org mode has a few properties for which inheritance is hard-coded, at least for the special applications for which they are used:

`COLUMNS`
: The `:COLUMNS:` property defines the format of column view (see Section 7.5 [Column view], page 67). It is inherited in the sense that the level where a `:COLUMNS:` property is defined is used as the starting point for a column view table, independently of the location in the subtree from where columns view is turned on.

`CATEGORY`
: For agenda view, a category set through a `:CATEGORY:` property applies to the entire subtree.

`ARCHIVE`
: For archiving, the `:ARCHIVE:` property may define the archive location for the entire subtree (see Section 9.6.1 [Moving subtrees], page 100).

`LOGGING`
: The LOGGING property may define logging settings for an entry or a subtree (see Section 5.3.2 [Tracking TODO state changes], page 52).

7.5 Column view

A great way to view and edit properties in an outline tree is *column view*. In column view, each outline node is turned into a table row. Columns in this table provide access to properties of the entries. Org mode implements columns by overlaying a tabular structure over the headline of each item. While the headlines have been turned into a table row, you can still change the visibility of the outline tree. For example, you get a compact table by switching to CONTENTS view (*S-TAB S-TAB*, or simply `c` while column view is active), but you can still open, read, and edit the entry below each headline. Or, you can switch to column view after executing a sparse tree command and in this way get a table only for the selected items. Column view also works in agenda buffers (see Chapter 10 [Agenda views], page 102) where queries have collected selected items, possibly from a number of files.

7.5.1 Defining columns

Setting up a column view first requires defining the columns. This is done by defining a column format line.

7.5.1.1 Scope of column definitions

To define a column format for an entire file, use a line like

```
#+COLUMNS: %25ITEM %TAGS %PRIORITY %TODO
```

To specify a format that only applies to a specific tree, add a `:COLUMNS:` property to the top node of that tree, for example:

```
** Top node for columns view
   :PROPERTIES:
   :COLUMNS: %25ITEM %TAGS %PRIORITY %TODO
   :END:
```

If a `:COLUMNS:` property is present in an entry, it defines columns for the entry itself, and for the entire subtree below it. Since the column definition is part of the hierarchical structure of the document, you can define columns on level 1 that are general enough for all sublevels, and more specific columns further down, when you edit a deeper part of the tree.

7.5.1.2 Column attributes

A column definition sets the attributes of a column. The general definition looks like this:

```
%[width]property[(title)][{summary-type}]
```

Except for the percent sign and the property name, all items are optional. The individual parts have the following meaning:

width — An integer specifying the width of the column in characters. If omitted, the width will be determined automatically.

property — The property that should be edited in this column. Special properties representing meta data are allowed here as well (see Section 7.2 [Special properties], page 66)

title — The header text for the column. If omitted, the property name is used.

{*summary-type*} — The summary type. If specified, the column values for parent nodes are computed from the children[1].
Supported summary types are:

`{+}` — Sum numbers in this column.

[1] If more than one summary type apply to the property, the parent values are computed according to the first of them.

`{+;%.1f}`	Like '+', but format result with '`%.1f`'.
`{$}`	Currency, short for '`+;%.2f`'.
`{min}`	Smallest number in column.
`{max}`	Largest number.
`{mean}`	Arithmetic mean of numbers.
`{X}`	Checkbox status, '`[X]`' if all children are '`[X]`'.
`{X/}`	Checkbox status, '`[n/m]`'.
`{X%}`	Checkbox status, '`[n%]`'.
`{:}`	Sum times, HH:MM, plain numbers are hours[2].
`{:min}`	Smallest time value in column.
`{:max}`	Largest time value.
`{:mean}`	Arithmetic mean of time values.
`{@min}`	Minimum age[3] (in days/hours/mins/seconds).
`{@max}`	Maximum age (in days/hours/mins/seconds).
`{@mean}`	Arithmetic mean of ages (in days/hours/mins/seconds).
`{est+}`	Add '`low-high`' estimates.

The `est+` summary type requires further explanation. It is used for combining estimates, expressed as '`low-high`' ranges or plain numbers. For example, instead of estimating a particular task will take 5 days, you might estimate it as 5–6 days if you're fairly confident you know how much work is required, or 1–10 days if you don't really know what needs to be done. Both ranges average at 5.5 days, but the first represents a more predictable delivery.

When combining a set of such estimates, simply adding the lows and highs produces an unrealistically wide result. Instead, `est+` adds the statistical mean and variance of the sub-tasks, generating a final estimate from the sum. For example, suppose you had ten tasks, each of which was estimated at 0.5 to 2 days of work. Straight addition produces an estimate of 5 to 20 days, representing what to expect if everything goes either extremely well or extremely poorly. In contrast, `est+` estimates the full job more realistically, at 10–15 days.

Numbers are right-aligned when a format specifier with an explicit width like `%5d` or `%5.1f` is used.

You can also define custom summary types by setting `org-columns-summary-types`, which see.

Here is an example for a complete columns definition, along with allowed values.

---

[2] A time can also be a duration, using effort modifiers defined in `org-effort-durations`, e.g., '`3d 1h`'. If any value in the column is as such, the summary will also be an effort duration.

[3] An age is defined as a duration since a given time-stamp (see Section 8.1 [Timestamps], page 73). It can also be expressed as days, hours, minutes and seconds, identified by '`d`', '`h`', '`m`' and '`s`' suffixes, all mandatory, e.g., '`0d 13h 0m 10s`'.

```
:COLUMNS:  %25ITEM %9Approved(Approved?){X} %Owner %11Status \
                   %10Time_Estimate{:} %CLOCKSUM %CLOCKSUM_T
:Owner_ALL:    Tammy Mark Karl Lisa Don
:Status_ALL:   "In progress" "Not started yet" "Finished" ""
:Approved_ALL: "[ ]" "[X]"
```

[4]

The first column, '%25ITEM', means the first 25 characters of the item itself, i.e., of the headline. You probably always should start the column definition with the 'ITEM' specifier. The other specifiers create columns 'Owner' with a list of names as allowed values, for 'Status' with four different possible values, and for a checkbox field 'Approved'. When no width is given after the '%' character, the column will be exactly as wide as it needs to be in order to fully display all values. The 'Approved' column does have a modified title ('Approved?', with a question mark). Summaries will be created for the 'Time_Estimate' column by adding time duration expressions like HH:MM, and for the 'Approved' column, by providing an '[X]' status if all children have been checked. The 'CLOCKSUM' and 'CLOCKSUM_T' columns are special, they lists the sums of CLOCK intervals in the subtree, either for all clocks or just for today.

7.5.2 Using column view

Turning column view on and off

C-c C-x C-c `org-columns`

Turn on column view. If the cursor is before the first headline in the file, or the function called with the universal prefix argument, column view is turned on for the entire file, using the `#+COLUMNS` definition. If the cursor is somewhere inside the outline, this command searches the hierarchy, up from point, for a `:COLUMNS:` property that defines a format. When one is found, the column view table is established for the tree starting at the entry that contains the `:COLUMNS:` property. If no such property is found, the format is taken from the `#+COLUMNS` line or from the variable `org-columns-default-format`, and column view is established for the current entry and its subtree.

r `org-columns-redo`

Recreate the column view, to include recent changes made in the buffer.

g `org-columns-redo`

Same as *r*.

q `org-columns-quit`

Exit column view.

Editing values

left right up down

Move through the column view from field to field.

S-left/right

Switch to the next/previous allowed value of the field. For this, you have to have specified allowed values for a property.

---

[4] Please note that the COLUMNS definition must be on a single line—it is wrapped here only because of formatting constraints.

1..9,0 Directly select the Nth allowed value, *0* selects the 10th value.

n `org-columns-next-allowed-value`
p `org-columns-previous-allowed-value`
: Same as *S-left/right*

e `org-columns-edit-value`
: Edit the property at point. For the special properties, this will invoke the same interface that you normally use to change that property. For example, when editing a TAGS property, the tag completion or fast selection interface will pop up.

C-c C-c `org-columns-set-tags-or-toggle`
: When there is a checkbox at point, toggle it.

v `org-columns-show-value`
: View the full value of this property. This is useful if the width of the column is smaller than that of the value.

`a` `org-columns-edit-allowed`
: Edit the list of allowed values for this property. If the list is found in the hierarchy, the modified value is stored there. If no list is found, the new value is stored in the first entry that is part of the current column view.

Modifying the table structure

`<` `org-columns-narrow`
`>` `org-columns-widen`
: Make the column narrower/wider by one character.

S-M-right `org-columns-new`
: Insert a new column, to the left of the current column.

S-M-left `org-columns-delete`
: Delete the current column.

7.5.3 Capturing column view

Since column view is just an overlay over a buffer, it cannot be exported or printed directly. If you want to capture a column view, use a `columnview` dynamic block (see Section A.7 [Dynamic blocks], page 247). The frame of this block looks like this:

```
* The column view
#+BEGIN: columnview :hlines 1 :id "label"

#+END:
```

This dynamic block has the following parameters:

`:id` This is the most important parameter. Column view is a feature that is often localized to a certain (sub)tree, and the capture block might be at a different location in the file. To identify the tree whose view to capture, you can use 4 values:

`local`	use the tree in which the capture block is located
`global`	make a global view, including all headings in the file

"file:*path-to-file*"
run column view at the top of this file

"*ID*" call column view in the tree that has an :ID: property with the value *label*. You can use *M-x org-id-copy RET* to create a globally unique ID for the current entry and copy it to the kill-ring.

:hlines When t, insert an hline after every line. When a number *N*, insert an hline before each headline with level <= *N*.

:vlines When set to t, force column groups to get vertical lines.

:maxlevel
When set to a number, don't capture entries below this level.

:skip-empty-rows
When set to t, skip rows where the only non-empty specifier of the column view is ITEM.

:indent When non-nil, indent each ITEM field according to its level.

The following commands insert or update the dynamic block:

C-c C-x i org-insert-columns-dblock
Insert a dynamic block capturing a column view. You will be prompted for the scope or ID of the view.

C-c C-c or *C-c C-x C-u* org-dblock-update
Update dynamic block at point.

C-u C-c C-x C-u org-update-all-dblocks
Update all dynamic blocks (see Section A.7 [Dynamic blocks], page 247). This is useful if you have several clock table blocks, column-capturing blocks or other dynamic blocks in a buffer.

You can add formulas to the column view table and you may add plotting instructions in front of the table—these will survive an update of the block. If there is a #+TBLFM: after the table, the table will actually be recalculated automatically after an update.

An alternative way to capture and process property values into a table is provided by Eric Schulte's org-collector.el which is a contributed package[5]. It provides a general API to collect properties from entries in a certain scope, and arbitrary Lisp expressions to process these values before inserting them into a table or a dynamic block.

7.6 The Property API

There is a full API for accessing and changing properties. This API can be used by Emacs Lisp programs to work with properties and to implement features based on them. For more information see Section A.11 [Using the property API], page 251.

---

[5] Contributed packages are not part of Emacs, but are distributed with the main distribution of Org (visit http://orgmode.org).

8 Dates and times

To assist project planning, TODO items can be labeled with a date and/or a time. The specially formatted string carrying the date and time information is called a *timestamp* in Org mode. This may be a little confusing because timestamp is often used to indicate when something was created or last changed. However, in Org mode this term is used in a much wider sense.

8.1 Timestamps, deadlines, and scheduling

A timestamp is a specification of a date (possibly with a time or a range of times) in a special format, either '`<2003-09-16 Tue>`'[1] or '`<2003-09-16 Tue 09:39>`' or '`<2003-09-16 Tue 12:00-12:30>`'[2]. A timestamp can appear anywhere in the headline or body of an Org tree entry. Its presence causes entries to be shown on specific dates in the agenda (see Section 10.3.1 [Weekly/daily agenda], page 104). We distinguish:

Plain timestamp; Event; Appointment

A simple timestamp just assigns a date/time to an item. This is just like writing down an appointment or event in a paper agenda. In the agenda display, the headline of an entry associated with a plain timestamp will be shown exactly on that date.

```
* Meet Peter at the movies
  <2006-11-01 Wed 19:15>
* Discussion on climate change
  <2006-11-02 Thu 20:00-22:00>
```

Timestamp with repeater interval

A timestamp may contain a *repeater interval*, indicating that it applies not only on the given date, but again and again after a certain interval of N days (d), weeks (w), months (m), or years (y). The following will show up in the agenda every Wednesday:

```
* Pick up Sam at school
  <2007-05-16 Wed 12:30 +1w>
```

Diary-style sexp entries

For more complex date specifications, Org mode supports using the special sexp diary entries implemented in the Emacs calendar/diary package[3]. For example with optional time

---

[1] In this simplest form, the day name is optional when you type the date yourself. However, any dates inserted or modified by Org will add that day name, for reading convenience.

[2] This is inspired by the standard ISO 8601 date/time format. To use an alternative format, see Section 8.2.2 [Custom time format], page 77.

[3] When working with the standard diary sexp functions, you need to be very careful with the order of the arguments. That order depends evilly on the variable `calendar-date-style` (or, for older Emacs versions, `european-calendar-style`). For example, to specify a date December 1, 2005, the call might look like `(diary-date 12 1 2005)` or `(diary-date 1 12 2005)` or `(diary-date 2005 12 1)`, depending on the settings. This has been the source of much confusion. Org mode users can resort to special versions of these functions like `org-date` or `org-anniversary`. These work just like the corresponding `diary-` functions, but with stable ISO order of arguments (year, month, day) wherever applicable, independent of the value of `calendar-date-style`.

```
* 22:00-23:00 The nerd meeting on every 2nd Thursday of the month
  <%%(diary-float t 4 2)>
```

Time/Date range

Two timestamps connected by '--' denote a range. The headline will be shown on the first and last day of the range, and on any dates that are displayed and fall in the range. Here is an example:

```
** Meeting in Amsterdam
   <2004-08-23 Mon>--<2004-08-26 Thu>
```

Inactive timestamp

Just like a plain timestamp, but with square brackets instead of angular ones. These timestamps are inactive in the sense that they do *not* trigger an entry to show up in the agenda.

```
* Gillian comes late for the fifth time
  [2006-11-01 Wed]
```

8.2 Creating timestamps

For Org mode to recognize timestamps, they need to be in the specific format. All commands listed below produce timestamps in the correct format.

C-c . `org-time-stamp`

Prompt for a date and insert a corresponding timestamp. When the cursor is at an existing timestamp in the buffer, the command is used to modify this timestamp instead of inserting a new one. When this command is used twice in succession, a time range is inserted.

C-c ! `org-time-stamp-inactive`

Like *C-c .*, but insert an inactive timestamp that will not cause an agenda entry.

C-u C-c .
C-u C-c ! Like *C-c .* and *C-c !*, but use the alternative format which contains date and time. The default time can be rounded to multiples of 5 minutes, see the option `org-time-stamp-rounding-minutes`.

C-c C-c Normalize timestamp, insert/fix day name if missing or wrong.

C-c < `org-date-from-calendar`

Insert a timestamp corresponding to the cursor date in the Calendar.

C-c > `org-goto-calendar`

Access the Emacs calendar for the current date. If there is a timestamp in the current line, go to the corresponding date instead.

C-c C-o `org-open-at-point`

Access the agenda for the date given by the timestamp or -range at point (see Section 10.3.1 [Weekly/daily agenda], page 104).

S-left `org-timestamp-down-day`
S-right `org-timestamp-up-day`

Change date at cursor by one day. These key bindings conflict with shift-selection and related modes (see Section 15.10.2 [Conflicts], page 238).

S-up `org-timestamp-up`
S-down `org-timestamp-down-down`
: Change the item under the cursor in a timestamp. The cursor can be on a year, month, day, hour or minute. When the timestamp contains a time range like '`15:30-16:30`', modifying the first time will also shift the second, shifting the time block with constant length. To change the length, modify the second time. Note that if the cursor is in a headline and not at a timestamp, these same keys modify the priority of an item. (see Section 5.4 [Priorities], page 55). The key bindings also conflict with shift-selection and related modes (see Section 15.10.2 [Conflicts], page 238).

C-c C-y `org-evaluate-time-range`
: Evaluate a time range by computing the difference between start and end. With a prefix argument, insert result after the time range (in a table: into the following column).

8.2.1 The date/time prompt

When Org mode prompts for a date/time, the default is shown in default date/time format, and the prompt therefore seems to ask for a specific format. But it will in fact accept date/time information in a variety of formats. Generally, the information should start at the beginning of the string. Org mode will find whatever information is in there and derive anything you have not specified from the *default date and time*. The default is usually the current date and time, but when modifying an existing timestamp, or when entering the second stamp of a range, it is taken from the stamp in the buffer. When filling in information, Org mode assumes that most of the time you will want to enter a date in the future: if you omit the month/year and the given day/month is *before* today, it will assume that you mean a future date[4]. If the date has been automatically shifted into the future, the time prompt will show this with '`(=>F)`.'

For example, let's assume that today is **June 13, 2006**. Here is how various inputs will be interpreted, the items filled in by Org mode are in **bold**.

`3-2-5`	⇒	`2003-02-05`
`2/5/3`	⇒	`2003-02-05`
`14`	⇒	**2006-06-**`14`
`12`	⇒	**2006-07-**`12`
`2/5`	⇒	**2007**`-02-05`
`Fri`	⇒	`nearest Friday after the default date`
`sep 15`	⇒	**2006**`-09-15`
`feb 15`	⇒	**2007**`-02-15`
`sep 12 9`	⇒	`2009-09-12`
`12:45`	⇒	**2006-06-13** `12:45`
`22 sept 0:34`	⇒	**2006**`-09-22 00:34`
`w4`	⇒	`ISO week four of the current year` **2006**
`2012 w4 fri`	⇒	`Friday of ISO week 4 in 2012`
`2012-w04-5`	⇒	`Same as above`

[4] See the variable `org-read-date-prefer-future`. You may set that variable to the symbol `time` to even make a time before now shift the date to tomorrow.

Furthermore you can specify a relative date by giving, as the *first* thing in the input: a plus/minus sign, a number and a letter ([hdwmy]) to indicate change in hours, days, weeks, months, or years. With a single plus or minus, the date is always relative to today. With a double plus or minus, it is relative to the default date. If instead of a single letter, you use the abbreviation of day name, the date will be the Nth such day, e.g.:

```
+0              ⇒ today
.               ⇒ today
+4d             ⇒ four days from today
+4              ⇒ same as above
+2w             ⇒ two weeks from today
++5             ⇒ five days from default date
+2tue           ⇒ second Tuesday from now
-wed            ⇒ last Wednesday
```

The function understands English month and weekday abbreviations. If you want to use unabbreviated names and/or other languages, configure the variables `parse-time-months` and `parse-time-weekdays`.

Not all dates can be represented in a given Emacs implementation. By default Org mode forces dates into the compatibility range 1970–2037 which works on all Emacs implementations. If you want to use dates outside of this range, read the docstring of the variable `org-read-date-force-compatible-dates`.

You can specify a time range by giving start and end times or by giving a start time and a duration (in HH:MM format). Use one or two dash(es) as the separator in the former case and use '+' as the separator in the latter case, e.g.:

```
11am-1:15pm     ⇒ 11:00-13:15
11am--1:15pm    ⇒ same as above
11am+2:15       ⇒ same as above
```

Parallel to the minibuffer prompt, a calendar is popped up[5]. When you exit the date prompt, either by clicking on a date in the calendar, or by pressing `RET`, the date selected in the calendar will be combined with the information entered at the prompt. You can control the calendar fully from the minibuffer:

`RET`	Choose date at cursor in calendar.
`mouse-1`	Select date by clicking on it.
`S-right/left`	One day forward/backward.
`S-down/up`	One week forward/backward.
`M-S-right/left`	One month forward/backward.
`> / <`	Scroll calendar forward/backward by one month.
`M-v / C-v`	Scroll calendar forward/backward by 3 months.
`M-S-down/up`	Scroll calendar forward/backward by one year.

The actions of the date/time prompt may seem complex, but I assure you they will grow on you, and you will start getting annoyed by pretty much any other way of entering a date/time out there. To help you understand what is going on, the current interpretation of your input will be displayed live in the minibuffer[6].

[5] If you don't need/want the calendar, configure the variable `org-popup-calendar-for-date-prompt`.

[6] If you find this distracting, turn the display off with `org-read-date-display-live`.

8.2.2 Custom time format

Org mode uses the standard ISO notation for dates and times as it is defined in ISO 8601. If you cannot get used to this and require another representation of date and time to keep you happy, you can get it by customizing the options `org-display-custom-times` and `org-time-stamp-custom-formats`.

C-c C-x C-t `org-toggle-time-stamp-overlays`
: Toggle the display of custom formats for dates and times.

Org mode needs the default format for scanning, so the custom date/time format does not *replace* the default format—instead it is put *over* the default format using text properties. This has the following consequences:

- You cannot place the cursor onto a timestamp anymore, only before or after.
- The *S-up/down* keys can no longer be used to adjust each component of a timestamp. If the cursor is at the beginning of the stamp, *S-up/down* will change the stamp by one day, just like *S-left/right*. At the end of the stamp, the time will be changed by one minute.
- If the timestamp contains a range of clock times or a repeater, these will not be overlaid, but remain in the buffer as they were.
- When you delete a timestamp character-by-character, it will only disappear from the buffer after *all* (invisible) characters belonging to the ISO timestamp have been removed.
- If the custom timestamp format is longer than the default and you are using dates in tables, table alignment will be messed up. If the custom format is shorter, things do work as expected.

8.3 Deadlines and scheduling

A timestamp may be preceded by special keywords to facilitate planning. Both the timestamp and the keyword have to be positioned immediately after the task they refer to.

DEADLINE
: Meaning: the task (most likely a TODO item, though not necessarily) is supposed to be finished on that date.

On the deadline date, the task will be listed in the agenda. In addition, the agenda for *today* will carry a warning about the approaching or missed deadline, starting `org-deadline-warning-days` before the due date, and continuing until the entry is marked DONE. An example:

```
*** TODO write article about the Earth for the Guide
    DEADLINE: <2004-02-29 Sun>
    The editor in charge is [[bbdb:Ford Prefect]]
```

You can specify a different lead time for warnings for a specific deadline using the following syntax. Here is an example with a warning period of 5 days `DEADLINE: <2004-02-29 Sun -5d>`. This warning is deactivated if the task gets scheduled and you set `org-agenda-skip-deadline-prewarning-if-scheduled` to `t`.

SCHEDULED

Meaning: you are planning to start working on that task on the given date.

The headline will be listed under the given date[7]. In addition, a reminder that the scheduled date has passed will be present in the compilation for *today*, until the entry is marked DONE, i.e., the task will automatically be forwarded until completed.

```
*** TODO Call Trillian for a date on New Years Eve.
    SCHEDULED: <2004-12-25 Sat>
```

If you want to *delay* the display of this task in the agenda, use `SCHEDULED: <2004-12-25 Sat -2d>`: the task is still scheduled on the 25th but will appear two days later. In case the task contains a repeater, the delay is considered to affect all occurrences; if you want the delay to only affect the first scheduled occurrence of the task, use `--2d` instead. See `org-scheduled-delay-days` and `org-agenda-skip-scheduled-delay-if-deadline` for details on how to control this globally or per agenda.

Important: Scheduling an item in Org mode should *not* be understood in the same way that we understand *scheduling a meeting*. Setting a date for a meeting is just a simple appointment, you should mark this entry with a simple plain timestamp, to get this item shown on the date where it applies. This is a frequent misunderstanding by Org users. In Org mode, *scheduling* means setting a date when you want to start working on an action item.

You may use timestamps with repeaters in scheduling and deadline entries. Org mode will issue early and late warnings based on the assumption that the timestamp represents the *nearest instance* of the repeater. However, the use of diary sexp entries like `<%%(diary-float t 42)>` in scheduling and deadline timestamps is limited. Org mode does not know enough about the internals of each sexp function to issue early and late warnings. However, it will show the item on each day where the sexp entry matches.

8.3.1 Inserting deadlines or schedules

The following commands allow you to quickly insert a deadline or to schedule an item:

C-c C-d `org-deadline`

Insert '`DEADLINE`' keyword along with a stamp. Any CLOSED timestamp will be removed. When called with a prefix arg, an existing deadline will be removed from the entry. Depending on the variable `org-log-redeadline`[8], a note will be taken when changing an existing deadline.

C-c C-s `org-schedule`

Insert '`SCHEDULED`' keyword along with a stamp. Any CLOSED timestamp will be removed. When called with a prefix argument, remove the scheduling date from the entry. Depending on the variable `org-log-reschedule`[9], a note will be taken when changing an existing scheduling time.

---

[7] It will still be listed on that date after it has been marked DONE. If you don't like this, set the variable `org-agenda-skip-scheduled-if-done`.

[8] with corresponding `#+STARTUP` keywords `logredeadline`, `lognoteredeadline`, and `nologredeadline`

[9] with corresponding `#+STARTUP` keywords `logreschedule`, `lognotereschedule`, and `nologreschedule`

`C-c / d` **org-check-deadlines**
: Create a sparse tree with all deadlines that are either past-due, or which will become due within `org-deadline-warning-days`. With *C-u* prefix, show all deadlines in the file. With a numeric prefix, check that many days. For example, *C-1 C-c / d* shows all deadlines due tomorrow.

`C-c / b` **org-check-before-date**
: Sparse tree for deadlines and scheduled items before a given date.

`C-c / a` **org-check-after-date**
: Sparse tree for deadlines and scheduled items after a given date.

Note that `org-schedule` and `org-deadline` supports setting the date by indicating a relative time: e.g., +1d will set the date to the next day after today, and –1w will set the date to the previous week before any current timestamp.

8.3.2 Repeated tasks

Some tasks need to be repeated again and again. Org mode helps to organize such tasks using a so-called repeater in a DEADLINE, SCHEDULED, or plain timestamp. In the following example

```
** TODO Pay the rent
   DEADLINE: <2005-10-01 Sat +1m>
```

the `+1m` is a repeater; the intended interpretation is that the task has a deadline on <2005-10-01> and repeats itself every (one) month starting from that time. You can use yearly, monthly, weekly, daily and hourly repeat cookies by using the `y/w/m/d/h` letters. If you need both a repeater and a special warning period in a deadline entry, the repeater should come first and the warning period last: `DEADLINE: <2005-10-01 Sat +1m -3d>`.

Deadlines and scheduled items produce entries in the agenda when they are over-due, so it is important to be able to mark such an entry as completed once you have done so. When you mark a DEADLINE or a SCHEDULE with the TODO keyword DONE, it will no longer produce entries in the agenda. The problem with this is, however, that then also the *next* instance of the repeated entry will not be active. Org mode deals with this in the following way: When you try to mark such an entry DONE (using *C-c C-t*), it will shift the base date of the repeating timestamp by the repeater interval, and immediately set the entry state back to TODO[10]. In the example above, setting the state to DONE would actually switch the date like this:

```
** TODO Pay the rent
   DEADLINE: <2005-11-01 Tue +1m>
```

To mark a task with a repeater as `DONE`, use *C-- 1 C-c C-t* (i.e., `org-todo` with a numeric prefix argument of -1.)

A timestamp[11] will be added under the deadline, to keep a record that you actually acted on the previous instance of this deadline.

---

[10] In fact, the target state is taken from, in this sequence, the `REPEAT_TO_STATE` property or the variable `org-todo-repeat-to-state`. If neither of these is specified, the target state defaults to the first state of the TODO state sequence.

[11] You can change this using the option `org-log-repeat`, or the `#+STARTUP` options `logrepeat`, `lognoterepeat`, and `nologrepeat`. With `lognoterepeat`, you will also be prompted for a note.

As a consequence of shifting the base date, this entry will no longer be visible in the agenda when checking past dates, but all future instances will be visible.

With the '+1m' cookie, the date shift will always be exactly one month. So if you have not paid the rent for three months, marking this entry DONE will still keep it as an overdue deadline. Depending on the task, this may not be the best way to handle it. For example, if you forgot to call your father for 3 weeks, it does not make sense to call him 3 times in a single day to make up for it. Finally, there are tasks like changing batteries which should always repeat a certain time *after* the last time you did it. For these tasks, Org mode has special repeaters '++' and '.+'. For example:

```
** TODO Call Father
   DEADLINE: <2008-02-10 Sun ++1w>
   Marking this DONE will shift the date by at least one week,
   but also by as many weeks as it takes to get this date into
   the future.  However, it stays on a Sunday, even if you called
   and marked it done on Saturday.
** TODO Empty kitchen trash
   DEADLINE: <2008-02-08 Fri 20:00 ++1d>
   Marking this DONE will shift the date by at least one day, and
   also by as many days as it takes to get the timestamp into the
   future.  Since there is a time in the timestamp, the next
   deadline in the future will be on today's date if you
   complete the task before 20:00.
** TODO Check the batteries in the smoke detectors
   DEADLINE: <2005-11-01 Tue .+1m>
   Marking this DONE will shift the date to one month after
   today.
```

You may have both scheduling and deadline information for a specific task. If the repeater is set for the scheduling information only, you probably want the repeater to be ignored after the deadline. If so, set the variable `org-agenda-skip-scheduled-if-deadline-is-shown` to `repeated-after-deadline`. However, any scheduling information without a repeater is no longer relevant once the task is done, and thus, removed upon repeating the task. If you want both scheduling and deadline information to repeat after the same interval, set the same repeater for both timestamps.

An alternative to using a repeater is to create a number of copies of a task subtree, with dates shifted in each copy. The command *C-c C-x c* was created for this purpose, it is described in Section 2.5 [Structure editing], page 9.

8.4 Clocking work time

Org mode allows you to clock the time you spend on specific tasks in a project. When you start working on an item, you can start the clock. When you stop working on that task, or when you mark the task done, the clock is stopped and the corresponding time interval is recorded. It also computes the total time spent on each subtree[12] of a project. And

[12] Clocking only works if all headings are indented with less than 30 stars. This is a hardcoded limitation of `lmax` in `org-clock-sum`.

it remembers a history or tasks recently clocked, so that you can jump quickly between a number of tasks absorbing your time.

To save the clock history across Emacs sessions, use

```
(setq org-clock-persist 'history)
(org-clock-persistence-insinuate)
```

When you clock into a new task after resuming Emacs, the incomplete clock[13] will be found (see Section 8.4.3 [Resolving idle time], page 85) and you will be prompted about what to do with it.

8.4.1 Clocking commands

C-c C-x C-i `org-clock-in`

Start the clock on the current item (clock-in). This inserts the CLOCK keyword together with a timestamp. If this is not the first clocking of this item, the multiple CLOCK lines will be wrapped into a `:LOGBOOK:` drawer (see also the variable `org-clock-into-drawer`). You can also overrule the setting of this variable for a subtree by setting a `CLOCK_INTO_DRAWER` or `LOG_INTO_DRAWER` property. When called with a *C-u* prefix argument, select the task from a list of recently clocked tasks. With two *C-u C-u* prefixes, clock into the task at point and mark it as the default task; the default task will then always be available with letter *d* when selecting a clocking task. With three *C-u C-u C-u* prefixes, force continuous clocking by starting the clock when the last clock stopped.

While the clock is running, the current clocking time is shown in the mode line, along with the title of the task. The clock time shown will be all time ever clocked for this task and its children. If the task has an effort estimate (see Section 8.5 [Effort estimates], page 86), the mode line displays the current clocking time against it[14] If the task is a repeating one (see Section 8.3.2 [Repeated tasks], page 79), only the time since the last reset of the task[15] will be shown. More control over what time is shown can be exercised with the `CLOCK_MODELINE_TOTAL` property. It may have the values `current` to show only the current clocking instance, `today` to show all time clocked on this task today (see also the variable `org-extend-today-until`), `all` to include all time, or `auto` which is the default[16].

Clicking with *mouse-1* onto the mode line entry will pop up a menu with clocking options.

C-c C-x C-o `org-clock-out`

Stop the clock (clock-out). This inserts another timestamp at the same location where the clock was last started. It also directly computes the resulting time and inserts it after the time range as '`=> HH:MM`'. See the variable `org-log-`

---

[13] To resume the clock under the assumption that you have worked on this task while outside Emacs, use `(setq org-clock-persist t)`.

[14] To add an effort estimate "on the fly", hook a function doing this to `org-clock-in-prepare-hook`.

[15] as recorded by the `LAST_REPEAT` property

[16] See also the variable `org-clock-modeline-total`.

`note-clock-out` for the possibility to record an additional note together with the clock-out timestamp[17].

C-c C-x C-x `org-clock-in-last`
: Reclock the last clocked task. With one *C-u* prefix argument, select the task from the clock history. With two *C-u* prefixes, force continuous clocking by starting the clock when the last clock stopped.

C-c C-x C-e `org-clock-modify-effort-estimate`
: Update the effort estimate for the current clock task.

C-c C-c or *C-c C-y* `org-evaluate-time-range`
: Recompute the time interval after changing one of the timestamps. This is only necessary if you edit the timestamps directly. If you change them with *S-cursor* keys, the update is automatic.

C-S-up/down `org-clock-timestamps-up/down`
: On `CLOCK` log lines, increase/decrease both timestamps so that the clock duration keeps the same.

S-M-up/down `org-timestamp-up/down`
: On `CLOCK` log lines, increase/decrease the timestamp at point and the one of the previous (or the next clock) timestamp by the same duration. For example, if you hit *S-M-up* to increase a clocked-out timestamp by five minutes, then the clocked-in timestamp of the next clock will be increased by five minutes.

C-c C-t `org-todo`
: Changing the TODO state of an item to DONE automatically stops the clock if it is running in this same item.

C-c C-x C-q `org-clock-cancel`
: Cancel the current clock. This is useful if a clock was started by mistake, or if you ended up working on something else.

C-c C-x C-j `org-clock-goto`
: Jump to the headline of the currently clocked in task. With a *C-u* prefix arg, select the target task from a list of recently clocked tasks.

C-c C-x C-d `org-clock-display`
: Display time summaries for each subtree in the current buffer. This puts overlays at the end of each headline, showing the total time recorded under that heading, including the time of any subheadings. You can use visibility cycling to study the tree, but the overlays disappear when you change the buffer (see variable `org-remove-highlights-with-change`) or press *C-c C-c*.

The *l* key may be used the agenda (see Section 10.3.1 [Weekly/daily agenda], page 104) to show which tasks have been worked on or closed during a day.

Important: note that both `org-clock-out` and `org-clock-in-last` can have a global key binding and will not modify the window disposition.

[17] The corresponding in-buffer setting is: `#+STARTUP: lognoteclock-out`

8.4.2 The clock table

Org mode can produce quite complex reports based on the time clocking information. Such a report is called a *clock table*, because it is formatted as one or several Org tables.

C-c C-x C-r `org-clock-report`

Insert a dynamic block (see Section A.7 [Dynamic blocks], page 247) containing a clock report as an Org mode table into the current file. When the cursor is at an existing clock table, just update it. When called with a prefix argument, jump to the first clock report in the current document and update it. The clock table always includes also trees with `:ARCHIVE:` tag.

C-c C-c or *C-c C-x C-u* `org-dblock-update`

Update dynamic block at point.

C-u C-c C-x C-u

Update all dynamic blocks (see Section A.7 [Dynamic blocks], page 247). This is useful if you have several clock table blocks in a buffer.

S-left
S-right `org-clocktable-try-shift`

Shift the current `:block` interval and update the table. The cursor needs to be in the `#+BEGIN: clocktable` line for this command. If `:block` is `today`, it will be shifted to `today-1` etc.

Here is an example of the frame for a clock table as it is inserted into the buffer with the *C-c C-x C-r* command:

```
#+BEGIN: clocktable :maxlevel 2 :emphasize nil :scope file
#+END: clocktable
```

The '`BEGIN`' line specifies a number of options to define the scope, structure, and formatting of the report. Defaults for all these options can be configured in the variable `org-clocktable-defaults`.

First there are options that determine which clock entries are to be selected:

`:maxlevel` Maximum level depth to which times are listed in the table. Clocks at deeper levels will be summed into the upper level.

`:scope` The scope to consider. This can be any of the following:

`nil`	the current buffer or narrowed region
`file`	the full current buffer
`subtree`	the subtree where the clocktable is located
`tree`*N*	the surrounding level N tree, for example `tree3`
`tree`	the surrounding level 1 tree
`agenda`	all agenda files
`("file"..)`	scan these files
`function`	the list of files returned by a function of no argument
`file-with-archives`	current file and its archives
`agenda-with-archives`	all agenda files, including archives

`:block` The time block to consider. This block is specified either absolutely, or relative to the current time and may be any of these formats:

`2007-12-31`	New year eve 2007
`2007-12`	December 2007
`2007-W50`	ISO-week 50 in 2007
`2007-Q2`	2nd quarter in 2007
`2007`	the year 2007
`today, yesterday, today-N`	a relative day
`thisweek, lastweek, thisweek-N`	a relative week
`thismonth, lastmonth, thismonth-N`	a relative month
`thisyear, lastyear, thisyear-N`	a relative year
`untilnow`	

Use *S-left/right* keys to shift the time interval.

`:tstart`
: A time string specifying when to start considering times. Relative times like "<-2w>" can also be used. See Section 10.3.3 [Matching tags and properties], page 107 for relative time syntax.

`:tend`
: A time string specifying when to stop considering times. Relative times like "<now>" can also be used. See Section 10.3.3 [Matching tags and properties], page 107 for relative time syntax.

`:wstart`
: The starting day of the week. The default is 1 for monday.

`:mstart`
: The starting day of the month. The default 1 is for the first day of the month.

`:step`
: `week` or `day`, to split the table into chunks. To use this, `:block` or `:tstart`, `:tend` are needed.

`:stepskip0`
: Do not show steps that have zero time.

`:fileskip0`
: Do not show table sections from files which did not contribute.

`:tags`
: A tags match to select entries that should contribute. See Section 10.3.3 [Matching tags and properties], page 107 for the match syntax.

Then there are options which determine the formatting of the table. These options are interpreted by the function `org-clocktable-write-default`, but you can specify your own function using the `:formatter` parameter.

`:emphasize`
: When `t`, emphasize level one and level two items.

`:lang`
: Language[18] to use for descriptive cells like "Task".

`:link`
: Link the item headlines in the table to their origins.

`:narrow`
: An integer to limit the width of the headline column in the org table. If you write it like '`50!`', then the headline will also be shortened in export.

`:indent`
: Indent each headline field according to its level.

`:tcolumns`
: Number of columns to be used for times. If this is smaller than `:maxlevel`, lower levels will be lumped into one column.

`:level`
: Should a level number column be included?

`:sort`
: A cons cell like containing the column to sort and a sorting type. E.g., `:sort (1 . ?a)` sorts the first column alphabetically.

`:compact`
: Abbreviation for `:level nil :indent t :narrow 40! :tcolumns 1` All are overwritten except if there is an explicit `:narrow`

`:timestamp`
: A timestamp for the entry, when available. Look for SCHEDULED, DEADLINE, TIMESTAMP and TIMESTAMP_IA, in this order.

---

[18] Language terms can be set through the variable `org-clock-clocktable-language-setup`.

`:properties`	List of properties that should be shown in the table. Each property will get its own column.
`:inherit-props`	When this flag is `t`, the values for `:properties` will be inherited.
`:formula`	Content of a `#+TBLFM` line to be added and evaluated. As a special case, '`:formula %`' adds a column with % time. If you do not specify a formula here, any existing formula below the clock table will survive updates and be evaluated.
`:formatter`	A function to format clock data and insert it into the buffer.

To get a clock summary of the current level 1 tree, for the current day, you could write

```
#+BEGIN: clocktable :maxlevel 2 :block today :scope tree1 :link t
#+END: clocktable
```

and to use a specific time range you could write[19]

```
#+BEGIN: clocktable :tstart "<2006-08-10 Thu 10:00>"
                    :tend "<2006-08-10 Thu 12:00>"
#+END: clocktable
```

A range starting a week ago and ending right now could be written as

```
#+BEGIN: clocktable :tstart "<-1w>" :tend "<now>"
#+END: clocktable
```

A summary of the current subtree with % times would be

```
#+BEGIN: clocktable :scope subtree :link t :formula %
#+END: clocktable
```

A horizontally compact representation of everything clocked during last week would be

```
#+BEGIN: clocktable :scope agenda :block lastweek :compact t
#+END: clocktable
```

8.4.3 Resolving idle time and continuous clocking

Resolving idle time

If you clock in on a work item, and then walk away from your computer—perhaps to take a phone call—you often need to "resolve" the time you were away by either subtracting it from the current clock, or applying it to another one.

By customizing the variable `org-clock-idle-time` to some integer, such as 10 or 15, Emacs can alert you when you get back to your computer after being idle for that many minutes[20], and ask what you want to do with the idle time. There will be a question waiting for you when you get back, indicating how much idle time has passed (constantly updated with the current amount), as well as a set of choices to correct the discrepancy:

---

[19] Note that all parameters must be specified in a single line—the line is broken here only to fit it into the manual.

[20] On computers using Mac OS X, idleness is based on actual user idleness, not just Emacs' idle time. For X11, you can install a utility program `x11idle.c`, available in the `contrib/scripts` directory of the Org git distribution, or install the `xprintidle` package and set it to the variable `org-clock-x11idle-program-name` if you are running Debian, to get the same general treatment of idleness. On other systems, idle time refers to Emacs idle time only.

k To keep some or all of the minutes and stay clocked in, press *k*. Org will ask how many of the minutes to keep. Press `RET` to keep them all, effectively changing nothing, or enter a number to keep that many minutes.

K If you use the shift key and press *K*, it will keep however many minutes you request and then immediately clock out of that task. If you keep all of the minutes, this is the same as just clocking out of the current task.

s To keep none of the minutes, use *s* to subtract all the away time from the clock, and then check back in from the moment you returned.

S To keep none of the minutes and just clock out at the start of the away time, use the shift key and press *S*. Remember that using shift will always leave you clocked out, no matter which option you choose.

C To cancel the clock altogether, use *C*. Note that if instead of canceling you subtract the away time, and the resulting clock amount is less than a minute, the clock will still be canceled rather than clutter up the log with an empty entry.

What if you subtracted those away minutes from the current clock, and now want to apply them to a new clock? Simply clock in to any task immediately after the subtraction. Org will notice that you have subtracted time "on the books", so to speak, and will ask if you want to apply those minutes to the next task you clock in on.

There is one other instance when this clock resolution magic occurs. Say you were clocked in and hacking away, and suddenly your cat chased a mouse who scared a hamster that crashed into your UPS's power button! You suddenly lose all your buffers, but thanks to auto-save you still have your recent Org mode changes, including your last clock in.

If you restart Emacs and clock into any task, Org will notice that you have a dangling clock which was never clocked out from your last session. Using that clock's starting time as the beginning of the unaccounted-for period, Org will ask how you want to resolve that time. The logic and behavior is identical to dealing with away time due to idleness; it is just happening due to a recovery event rather than a set amount of idle time.

You can also check all the files visited by your Org agenda for dangling clocks at any time using *M-x org-resolve-clocks RET* (or *C-c C-x C-z*).

Continuous clocking

You may want to start clocking from the time when you clocked out the previous task. To enable this systematically, set `org-clock-continuously` to `t`. Each time you clock in, Org retrieves the clock-out time of the last clocked entry for this session, and start the new clock from there.

If you only want this from time to time, use three universal prefix arguments with `org-clock-in` and two *C-u C-u* with `org-clock-in-last`.

8.5 Effort estimates

If you want to plan your work in a very detailed way, or if you need to produce offers with quotations of the estimated work effort, you may want to assign effort estimates to entries. If you are also clocking your work, you may later want to compare the planned effort with the actual working time, a great way to improve planning estimates. Effort estimates are stored in a special property `EFFORT`. You can set the effort for an entry with the following commands:

C-c C-x e `org-set-effort`
: Set the effort estimate for the current entry. With a numeric prefix argument, set it to the Nth allowed value (see below). This command is also accessible from the agenda with the `e` key.

C-c C-x C-e `org-clock-modify-effort-estimate`
: Modify the effort estimate of the item currently being clocked.

Clearly the best way to work with effort estimates is through column view (see Section 7.5 [Column view], page 67). You should start by setting up discrete values for effort estimates, and a `COLUMNS` format that displays these values together with clock sums (if you want to clock your time). For a specific buffer you can use

```
#+PROPERTY: Effort_ALL 0 0:10 0:30 1:00 2:00 3:00 4:00 5:00 6:00 7:00
#+COLUMNS: %40ITEM(Task) %17Effort(Estimated Effort){:} %CLOCKSUM
```

or, even better, you can set up these values globally by customizing the variables `org-global-properties` and `org-columns-default-format`. In particular if you want to use this setup also in the agenda, a global setup may be advised.

The way to assign estimates to individual items is then to switch to column mode, and to use *S-right* and *S-left* to change the value. The values you enter will immediately be summed up in the hierarchy. In the column next to it, any clocked time will be displayed.

If you switch to column view in the daily/weekly agenda, the effort column will summarize the estimated work effort for each day[21], and you can use this to find space in your schedule. To get an overview of the entire part of the day that is committed, you can set the option `org-agenda-columns-add-appointments-to-effort-sum`. The appointments on a day that take place over a specified time interval will then also be added to the load estimate of the day.

Effort estimates can be used in secondary agenda filtering that is triggered with the `/` key in the agenda (see Section 10.5 [Agenda commands], page 115). If you have these estimates defined consistently, two or three key presses will narrow down the list to stuff that fits into an available time slot.

8.6 Taking notes with a timer

Org provides two types of timers. There is a relative timer that counts up, which can be useful when taking notes during, for example, a meeting or a video viewing. There is also a countdown timer.

The relative and countdown are started with separate commands.

C-c C-x 0 `org-timer-start`
: Start or reset the relative timer. By default, the timer is set to 0. When called with a *C-u* prefix, prompt the user for a starting offset. If there is a timer string at point, this is taken as the default, providing a convenient way

[21] Please note the pitfalls of summing hierarchical data in a flat list (see Section 10.8 [Agenda column view], page 128).

to restart taking notes after a break in the process. When called with a double prefix argument *C-u C-u*, change all timer strings in the active region by a certain amount. This can be used to fix timer strings if the timer was not started at exactly the right moment.

C-c C-x ; `org-timer-set-timer`
: Start a countdown timer. The user is prompted for a duration. `org-timer-default-timer` sets the default countdown value. Giving a numeric prefix argument overrides this default value. This command is available as *;* in agenda buffers.

Once started, relative and countdown timers are controlled with the same commands.

C-c C-x . `org-timer`
: Insert the value of the current relative or countdown timer into the buffer. If no timer is running, the relative timer will be started. When called with a prefix argument, the relative timer is restarted.

C-c C-x - `org-timer-item`
: Insert a description list item with the value of the current relative or countdown timer. With a prefix argument, first reset the relative timer to 0.

M-RET `org-insert-heading`
: Once the timer list is started, you can also use *M-RET* to insert new timer items.

C-c C-x , `org-timer-pause-or-continue`
: Pause the timer, or continue it if it is already paused.

C-c C-x _ `org-timer-stop`
: Stop the timer. After this, you can only start a new timer, not continue the old one. This command also removes the timer from the mode line.

9 Capture - Refile - Archive

An important part of any organization system is the ability to quickly capture new ideas and tasks, and to associate reference material with them. Org does this using a process called *capture*. It also can store files related to a task (*attachments*) in a special directory. Once in the system, tasks and projects need to be moved around. Moving completed project trees to an archive file keeps the system compact and fast.

9.1 Capture

Capture lets you quickly store notes with little interruption of your work flow. Org's method for capturing new items is heavily inspired by John Wiegley excellent `remember.el` package. Up to version 6.36, Org used a special setup for `remember.el`, then replaced it with `org-remember.el`. As of version 8.0, `org-remember.el` has been completely replaced by `org-capture.el`.

If your configuration depends on `org-remember.el`, you need to update it and use the setup described below. To convert your `org-remember-templates`, run the command

M-x org-capture-import-remember-templates RET

and then customize the new variable with *M-x customize-variable org-capture-templates*, check the result, and save the customization.

9.1.1 Setting up capture

The following customization sets a default target file for notes, and defines a global key[1] for capturing new material.

```
(setq org-default-notes-file (concat org-directory "/notes.org"))
(define-key global-map "\C-cc" 'org-capture)
```

9.1.2 Using capture

C-c c `org-capture`

Call the command `org-capture`. Note that this key binding is global and not active by default: you need to install it. If you have templates defined see Section 9.1.3 [Capture templates], page 90, it will offer these templates for selection or use a new Org outline node as the default template. It will insert the template into the target file and switch to an indirect buffer narrowed to this new node. You may then insert the information you want.

C-c C-c `org-capture-finalize`

Once you have finished entering information into the capture buffer, *C-c C-c* will return you to the window configuration before the capture process, so that you can resume your work without further distraction. When called with a prefix arg, finalize and then jump to the captured item.

C-c C-w `org-capture-refile`

Finalize the capture process by refiling (see Section 9.5 [Refile and copy], page 99) the note to a different place. Please realize that this is a normal

[1] Please select your own key, *C-c c* is only a suggestion.

refiling command that will be executed—so the cursor position at the moment you run this command is important. If you have inserted a tree with a parent and children, first move the cursor back to the parent. Any prefix argument given to this command will be passed on to the `org-refile` command.

C-c C-k `org-capture-kill`
: Abort the capture process and return to the previous state.

You can also call `org-capture` in a special way from the agenda, using the *k c* key combination. With this access, any timestamps inserted by the selected capture template will default to the cursor date in the agenda, rather than to the current date.

To find the locations of the last stored capture, use `org-capture` with prefix commands:

C-u C-c c
: Visit the target location of a capture template. You get to select the template in the usual way.

C-u C-u C-c c
: Visit the last stored capture item in its buffer.

You can also jump to the bookmark `org-capture-last-stored`, which will automatically be created unless you set `org-capture-bookmark` to `nil`.

To insert the capture at point in an Org buffer, call `org-capture` with a `C-0` prefix argument.

9.1.3 Capture templates

You can use templates for different types of capture items, and for different target locations. The easiest way to create such templates is through the customize interface.

C-c c C Customize the variable `org-capture-templates`.

Before we give the formal description of template definitions, let's look at an example. Say you would like to use one template to create general TODO entries, and you want to put these entries under the heading '`Tasks`' in your file `~/org/gtd.org`. Also, a date tree in the file `journal.org` should capture journal entries. A possible configuration would look like:

```
(setq org-capture-templates
 '(("t" "Todo" entry (file+headline "~/org/gtd.org" "Tasks")
        "* TODO %?\n  %i\n  %a")
   ("j" "Journal" entry (file+olp+datetree "~/org/journal.org")
        "* %?\nEntered on %U\n  %i\n  %a")))
```

If you then press *C-c c t*, Org will prepare the template for you like this:

```
* TODO
  [[file:link to where you initiated capture]]
```

During expansion of the template, `%a` has been replaced by a link to the location from where you called the capture command. This can be extremely useful for deriving tasks from emails, for example. You fill in the task definition, press *C-c C-c* and Org returns you to the same place where you started the capture process.

To define special keys to capture to a particular template without going through the interactive template selection, you can create your key binding like this:

```
(define-key global-map "\C-cx"
  (lambda () (interactive) (org-capture nil "x")))
```

9.1.3.1 Template elements

Now lets look at the elements of a template definition. Each entry in `org-capture-templates` is a list with the following items:

keys The keys that will select the template, as a string, characters only, for example `"a"` for a template to be selected with a single key, or `"bt"` for selection with two keys. When using several keys, keys using the same prefix key must be sequential in the list and preceded by a 2-element entry explaining the prefix key, for example

```
("b" "Templates for marking stuff to buy")
```

If you do not define a template for the *C* key, this key will be used to open the customize buffer for this complex variable.

description

A short string describing the template, which will be shown during selection.

type The type of entry, a symbol. Valid values are:

`entry` An Org mode node, with a headline. Will be filed as the child of the target entry or as a top-level entry. The target file should be an Org mode file.

`item` A plain list item, placed in the first plain list at the target location. Again the target file should be an Org file.

`checkitem`

A checkbox item. This only differs from the plain list item by the default template.

`table-line`

a new line in the first table at the target location. Where exactly the line will be inserted depends on the properties `:prepend` and `:table-line-pos` (see below).

`plain` Text to be inserted as it is.

target Specification of where the captured item should be placed. In Org mode files, targets usually define a node. Entries will become children of this node. Other types will be added to the table or list in the body of this node. Most target specifications contain a file name. If that file name is the empty string, it defaults to `org-default-notes-file`. A file can also be given as a variable or as a function called with no argument. When an absolute path is not specified for a target, it is taken as relative to `org-directory`.

Valid values are:

`(file "path/to/file")`

Text will be placed at the beginning or end of that file.

`(id "id of existing org entry")`

Filing as child of this entry, or in the body of the entry.

`(file+headline "path/to/file" "node headline")`
: Fast configuration if the target heading is unique in the file.

`(file+olp "path/to/file" "Level 1 heading" "Level 2" ...)`
: For non-unique headings, the full path is safer.

`(file+regexp "path/to/file" "regexp to find location")`
: Use a regular expression to position the cursor.

`(file+olp+datetree "path/to/file" [ "Level 1 heading" ....])`
: This target[2] will create a heading in a date tree[3] for today's date. If the optional outline path is given, the tree will be built under the node it is pointing to, instead of at top level. Check out the `:time-prompt` and `:tree-type` properties below for additional options.

`(file+function "path/to/file" function-finding-location)`
: A function to find the right location in the file.

`(clock)`
: File to the entry that is currently being clocked.

`(function function-finding-location)`
: Most general way: write your own function which both visits the file and moves point to the right location.

template The template for creating the capture item. If you leave this empty, an appropriate default template will be used. Otherwise this is a string with escape codes, which will be replaced depending on time and context of the capture call. The string with escapes may be loaded from a template file, using the special syntax `(file "path/to/template")`. See below for more details.

properties The rest of the entry is a property list of additional options. Recognized properties are:

`:prepend`
: Normally new captured information will be appended at the target location (last child, last table line, last list item...). Setting this property will change that.

`:immediate-finish`
: When set, do not offer to edit the information, just file it away immediately. This makes sense if the template only needs information that can be added automatically.

`:empty-lines`
: Set this to the number of lines to insert before and after the new item. Default 0, only common other value is 1.

`:clock-in`
: Start the clock in this item.

---

[2] Org used to offer four different targets for date/week tree capture. Now, Org automatically translates these to use `file+olp+datetree`, applying the `:time-prompt` and `:tree-type` properties. Please rewrite your date/week-tree targets using `file+olp+datetree` since the older targets are now deprecated.

[3] A date tree is an outline structure with years on the highest level, months or ISO-weeks as sublevels and then dates on the lowest level. Tags are allowed in the tree structure.

`:clock-keep`
: Keep the clock running when filing the captured entry.

`:clock-resume`
: If starting the capture interrupted a clock, restart that clock when finished with the capture. Note that `:clock-keep` has precedence over `:clock-resume`. When setting both to `t`, the current clock will run and the previous one will not be resumed.

`:time-prompt`
: Prompt for a date/time to be used for date/week trees and when filling the template. Without this property, capture uses the current date and time. Even if this property has not been set, you can force the same behavior by calling `org-capture` with a *C-1* prefix argument.

`:tree-type`
: When 'week', make a week tree instead of the month tree, i.e. place the headings for each day under a heading with the current iso week.

`:unnarrowed`
: Do not narrow the target buffer, simply show the full buffer. Default is to narrow it so that you only see the new material.

`:table-line-pos`
: Specification of the location in the table where the new line should be inserted. It can be a string, a variable holding a string or a function returning a string. The string should look like `"II-3"` meaning that the new line should become the third line before the second horizontal separator line.

`:kill-buffer`
: If the target file was not yet visited when capture was invoked, kill the buffer again after capture is completed.

9.1.3.2 Template expansion

In the template itself, special `%`-escapes[4] allow dynamic insertion of content. The templates are expanded in the order given here:

`%[file]`	Insert the contents of the file given by *file*.
`%(sexp)`	Evaluate Elisp *sexp* and replace with the result. For convenience, %:keyword (see below) placeholders within the expression will be expanded prior to this. The sexp must return a string.
`%<...>`	The result of format-time-string on the ... format specification.
`%t`	Timestamp, date only.
`%T`	Timestamp, with date and time.
`%u, %U`	Like the above, but inactive timestamps.
`%i`	Initial content, the region when capture is called while the region is active. The entire text will be indented like `%i` itself.
`%a`	Annotation, normally the link created with `org-store-link`.

[4] If you need one of these sequences literally, escape the `%` with a backslash.

`%A`	Like `%a`, but prompt for the description part.
`%l`	Like %a, but only insert the literal link.
`%c`	Current kill ring head.
`%x`	Content of the X clipboard.
`%k`	Title of the currently clocked task.
`%K`	Link to the currently clocked task.
`%n`	User name (taken from `user-full-name`).
`%f`	File visited by current buffer when org-capture was called.
`%F`	Full path of the file or directory visited by current buffer.
`%:keyword`	Specific information for certain link types, see below.
`%^g`	Prompt for tags, with completion on tags in target file.
`%^G`	Prompt for tags, with completion all tags in all agenda files.
`%^t`	Like `%t`, but prompt for date. Similarly `%^T`, `%^u`, `%^U`. You may define a prompt like `%^{Birthday}t`.
`%^C`	Interactive selection of which kill or clip to use.
`%^L`	Like `%^C`, but insert as link.
`%^{prop}p`	Prompt the user for a value for property *prop*.
`%^{prompt}`	prompt the user for a string and replace this sequence with it. You may specify a default value and a completion table with %^{prompt|default|completion2|completion3...}. The arrow keys access a prompt-specific history.
`%\1 ... %\N`	Insert the text entered at the Nth %^{*prompt*}, where `N` is a number, starting from 1.[5]
`%?`	After completing the template, position cursor here.

For specific link types, the following keywords will be defined[6]:

```
Link type                              |  Available keywords
---------------------------------------+------------------------------------------
bbdb                                   |  %:name %:company
irc                                    |  %:server %:port %:nick
vm, vm-imap, wl, mh, mew, rmail,       |  %:type %:subject %:message-id
gnus, notmuch                          |  %:from %:fromname %:fromaddress
                                       |  %:to    %:toname    %:toaddress
                                       |  %:date (message date header field)
                                       |  %:date-timestamp (date as active timestamp)
                                       |  %:date-timestamp-inactive (date as inactive timestamp)
                                       |  %:fromto (either "to NAME" or "from NAME")[7]
gnus                                   |  %:group, for messages also all email fields
eww, w3, w3m                           |  %:url
info                                   |  %:file %:node
calendar                               |  %:date
org-protocol                           |  %:link %:description %:annotation
```

To place the cursor after template expansion use:

`%?`	After completing the template, position cursor here.

9.1.3.3 Templates in contexts

To control whether a capture template should be accessible from a specific context, you can customize `org-capture-templates-contexts`. Let's say for example that you have a

[5] As required in Emacs Lisp, it is necessary to escape any backslash character in a string with another backslash. So, in order to use '`%\1`' placeholder, you need to write '`%\\1`' in the template.

[6] If you define your own link types (see Section A.3 [Adding hyperlink types], page 241), any property you store with `org-store-link-props` can be accessed in capture templates in a similar way.

[7] This will always be the other, not the user. See the variable `org-from-is-user-regexp`.

capture template "p" for storing Gnus emails containing patches. Then you would configure this option like this:

```
(setq org-capture-templates-contexts
      '(("p" (in-mode . "message-mode"))))
```

You can also tell that the command key "p" should refer to another template. In that case, add this command key like this:

```
(setq org-capture-templates-contexts
      '(("p" "q" (in-mode . "message-mode"))))
```

See the docstring of the variable for more information.

9.2 Attachments

It is often useful to associate reference material with an outline node/task. Small chunks of plain text can simply be stored in the subtree of a project. Hyperlinks (see Chapter 4 [Hyperlinks], page 38) can establish associations with files that live elsewhere on your computer or in the cloud, like emails or source code files belonging to a project. Another method is *attachments*, which are files located in a directory belonging to an outline node. Org uses directories named by the unique ID of each entry. These directories are located in the `data` directory which lives in the same directory where your Org file lives[8]. If you initialize this directory with `git init`, Org will automatically commit changes when it sees them. The attachment system has been contributed to Org by John Wiegley.

In cases where it seems better to do so, you can also attach a directory of your choice to an entry. You can also make children inherit the attachment directory from a parent, so that an entire subtree uses the same attached directory.

The following commands deal with attachments:

C-c C-a `org-attach`
: The dispatcher for commands related to the attachment system. After these keys, a list of commands is displayed and you must press an additional key to select a command:

  `a` `org-attach-attach`
  : Select a file and move it into the task's attachment directory. The file will be copied, moved, or linked, depending on `org-attach-method`. Note that hard links are not supported on all systems.

  c/m/l
  : Attach a file using the copy/move/link method. Note that hard links are not supported on all systems.

  u `org-attach-url`
  : Attach a file from URL

  n `org-attach-new`
  : Create a new attachment as an Emacs buffer.

  z `org-attach-sync`
  : Synchronize the current task with its attachment directory, in case you added attachments yourself.

[8] If you move entries or Org files from one directory to another, you may want to configure `org-attach-directory` to contain an absolute path.

o `org-attach-open`
Open current task's attachment. If there is more than one, prompt for a file name first. Opening will follow the rules set by `org-file-apps`. For more details, see the information on following hyperlinks (see Section 4.4 [Handling links], page 41).

O `org-attach-open-in-emacs`
Also open the attachment, but force opening the file in Emacs.

f `org-attach-reveal`
Open the current task's attachment directory.

F `org-attach-reveal-in-emacs`
Also open the directory, but force using `dired` in Emacs.

d `org-attach-delete-one`
Select and delete a single attachment.

D `org-attach-delete-all`
Delete all of a task's attachments. A safer way is to open the directory in `dired` and delete from there.

s `org-attach-set-directory`
Set a specific directory as the entry's attachment directory. This works by putting the directory path into the `ATTACH_DIR` property.

i `org-attach-set-inherit`
Set the `ATTACH_DIR_INHERIT` property, so that children will use the same directory for attachments as the parent does.

9.3 RSS feeds

Org can add and change entries based on information found in RSS feeds and Atom feeds. You could use this to make a task out of each new podcast in a podcast feed. Or you could use a phone-based note-creating service on the web to import tasks into Org. To access feeds, configure the variable `org-feed-alist`. The docstring of this variable has detailed information. Here is just an example:

```
(setq org-feed-alist
     '(("Slashdot"
        "http://rss.slashdot.org/Slashdot/slashdot"
        "~/txt/org/feeds.org" "Slashdot Entries")))
```

will configure that new items from the feed provided by `rss.slashdot.org` will result in new entries in the file `~/org/feeds.org` under the heading '`Slashdot Entries`', whenever the following command is used:

C-c C-x g `org-feed-update-all`
C-c C-x g Collect items from the feeds configured in `org-feed-alist` and act upon them.

C-c C-x G `org-feed-goto-inbox`
Prompt for a feed name and go to the inbox configured for this feed.

Under the same headline, Org will create a drawer 'FEEDSTATUS' in which it will store information about the status of items in the feed, to avoid adding the same item several times.

For more information, including how to read atom feeds, see `org-feed.el` and the docstring of `org-feed-alist`.

9.4 Protocols for external access

Org protocol is a mean to trigger custom actions in Emacs from external applications. Any application that supports calling external programs with an URL as argument may be used with this functionality. For example, you can configure bookmarks in your web browser to send a link to the current page to Org and create a note from it using capture (see Section 9.1 [Capture], page 89). You can also create a bookmark that tells Emacs to open the local source file of a remote website you are browsing.

In order to use Org protocol from an application, you need to register 'org-protocol://' as a valid scheme-handler. External calls are passed to Emacs through the `emacsclient` command, so you also need to ensure an Emacs server is running. More precisely, when the application calls

```
emacsclient org-protocol://PROTOCOL?key1=val1&key2=val2
```

Emacs calls the handler associated to 'PROTOCOL' with argument '(:key1 val1 :key2 val2)'.

Org protocol comes with three predefined protocols, detailed in the following sections. Configure `org-protocol-protocol-alist` to define your own.

9.4.1 store-link protocol

Using `store-link` handler, you can copy links, insertable through *M-x org-insert-link* or yanking thereafter. More precisely, the command

```
emacsclient org-protocol://store-link?url=URL&title=TITLE
```

stores the following link:

```
[[URL][TITLE]]
```

In addition, 'URL' is pushed on the kill-ring for yanking. You need to encode 'URL' and 'TITLE' if they contain slashes, and probably quote those for the shell.

To use this feature from a browser, add a bookmark with an arbitrary name, e.g., 'Org: store-link' and enter this as *Location*:

```
javascript:location.href='org-protocol://store-link?url='+
      encodeURIComponent(location.href);
```

9.4.2 capture protocol

Activating `capture` handler pops up a 'Capture' buffer and fills the capture template associated to the 'X' key with them.

```
emacsclient org-protocol://capture?template=X?url=URL?title=TITLE?body=BODY
```

To use this feature, add a bookmark with an arbitrary name, e.g. 'Org: capture' and enter this as 'Location':

```
javascript:location.href='org-protocol://template=x'+
      '&url='+encodeURIComponent(window.location.href)+
      '&title='+encodeURIComponent(document.title)+
      '&body='+encodeURIComponent(window.getSelection());
```

The result depends on the capture template used, which is set in the bookmark itself, as in the example above, or in `org-protocol-default-template-key`.

The following template placeholders are available:

```
%:link          The URL
%:description   The webpage title
%:annotation    Equivalent to [[%:link][%:description]]
%i              The selected text
```

9.4.3 open-source protocol

The `open-source` handler is designed to help with editing local sources when reading a document. To that effect, you can use a bookmark with the following location:

```
javascript:location.href='org-protocol://open-source?&url='+
      encodeURIComponent(location.href)
```

The variable `org-protocol-project-alist` maps URLs to local file names, by stripping URL parameters from the end and replacing the `:base-url` with `:working-directory` and `:online-suffix` with `:working-suffix`. For example, assuming you own a local copy of http://orgmode.org/worg/ contents at `/home/user/worg`, you can set `org-protocol-project-alist` to the following

```
(setq org-protocol-project-alist
      '(("Worg"
 :base-url "http://orgmode.org/worg/"
 :working-directory "/home/user/worg/"
 :online-suffix ".html"
 :working-suffix ".org")))
```

If you are now browsing http://orgmode.org/worg/org-contrib/org-protocol.html and find a typo or have an idea about how to enhance the documentation, simply click the bookmark and start editing.

However, such mapping may not yield the desired results. Suppose you maintain an online store located at http://example.com/. The local sources reside in `/home/user/example/`. It is common practice to serve all products in such a store through one file and rewrite URLs that do not match an existing file on the server. That way, a request to http://example.com/print/posters.html might be rewritten on the server to something like http://example.com/shop/products.php/posters.html.php. The `open-source` handler probably cannot find a file named `/home/user/example/print/posters.html.php` and fails.

Such an entry in `org-protocol-project-alist` may hold an additional property `:rewrites`. This property is a list of cons cells, each of which maps a regular expression to a path relative to the `:working-directory`.

Now map the URL to the path `/home/user/example/products.php` by adding `:rewrites` rules like this:

```
(setq org-protocol-project-alist
      '(("example.com"
         :base-url "http://example.com/"
         :working-directory "/home/user/example/"
         :online-suffix ".php"
         :working-suffix ".php"
         :rewrites (("example.com/print/" . "products.php")
                    ("example.com/$" . "index.php")))))
```

Since 'example.com/$' is used as a regular expression, it maps http://example.com/, https://example.com, http://www.example.com/ and similar to /home/user/example/index.php.

The :rewrites rules are searched as a last resort if and only if no existing file name is matched.

Two functions can help you filling org-protocol-project-alist with valid contents: org-protocol-create and org-protocol-create-for-org. The latter is of use if you're editing an Org file that is part of a publishing project.

9.5 Refile and copy

When reviewing the captured data, you may want to refile or to copy some of the entries into a different list, for example into a project. Cutting, finding the right location, and then pasting the note is cumbersome. To simplify this process, you can use the following special command:

C-c M-w org-copy
: Copying works like refiling, except that the original note is not deleted.

C-c C-w org-refile
: Refile the entry or region at point. This command offers possible locations for refiling the entry and lets you select one with completion. The item (or all items in the region) is filed below the target heading as a subitem. Depending on org-reverse-note-order, it will be either the first or last subitem.
By default, all level 1 headlines in the current buffer are considered to be targets, but you can have more complex definitions across a number of files. See the variable org-refile-targets for details. If you would like to select a location via a file-path-like completion along the outline path, see the variables org-refile-use-outline-path and org-outline-path-complete-in-steps. If you would like to be able to create new nodes as new parents for refiling on the fly, check the variable org-refile-allow-creating-parent-nodes. When the variable org-log-refile[9] is set, a timestamp or a note will be recorded when an entry has been refiled.

C-u C-c C-w
: Use the refile interface to jump to a heading.

C-u C-u C-c C-w org-refile-goto-last-stored
: Jump to the location where org-refile last moved a tree to.

[9] with corresponding #+STARTUP keywords logrefile, lognoterefile, and nologrefile

C-2 C-c C-w
: Refile as the child of the item currently being clocked.

C-3 C-c C-w
: Refile and keep the entry in place. Also see `org-refile-keep` to make this the default behavior, and beware that this may result in duplicated `ID` properties.

C-0 C-c C-w or *C-u C-u C-u C-c C-w* `org-refile-cache-clear`
: Clear the target cache. Caching of refile targets can be turned on by setting `org-refile-use-cache`. To make the command see new possible targets, you have to clear the cache with this command.

9.6 Archiving

When a project represented by a (sub)tree is finished, you may want to move the tree out of the way and to stop it from contributing to the agenda. Archiving is important to keep your working files compact and global searches like the construction of agenda views fast.

C-c C-x C-a `org-archive-subtree-default`
: Archive the current entry using the command specified in the variable `org-archive-default-command`.

9.6.1 Moving a tree to the archive file

The most common archiving action is to move a project tree to another file, the archive file.

C-c C-x C-s or short *C-c $* `org-archive-subtree`
: Archive the subtree starting at the cursor position to the location given by `org-archive-location`.

C-u C-c C-x C-s
: Check if any direct children of the current headline could be moved to the archive. To do this, each subtree is checked for open TODO entries. If none are found, the command offers to move it to the archive location. If the cursor is *not* on a headline when this command is invoked, the level 1 trees will be checked.

C-u C-u C-c C-x C-s
: As above, but check subtree for timestamps instead of TODO entries. The command will offer to archive the subtree if it *does* contain a timestamp, and that timestamp is in the past.

The default archive location is a file in the same directory as the current file, with the name derived by appending `_archive` to the current file name. You can also choose what heading to file archived items under, with the possibility to add them to a datetree in a file. For information and examples on how to specify the file and the heading, see the documentation string of the variable `org-archive-location`.

There is also an in-buffer option for setting this variable, for example:

```
#+ARCHIVE: %s_done::
```

If you would like to have a special ARCHIVE location for a single entry or a (sub)tree, give the entry an `:ARCHIVE:` property with the location as the value (see Chapter 7 [Properties and columns], page 64).

When a subtree is moved, it receives a number of special properties that record context information like the file from where the entry came, its outline path the archiving time etc. Configure the variable `org-archive-save-context-info` to adjust the amount of information added.

9.6.2 Internal archiving

If you want to just switch off—for agenda views—certain subtrees without moving them to a different file, you can use the archive tag.

A headline that is marked with the '`:ARCHIVE:`' tag (see Chapter 6 [Tags], page 59) stays at its location in the outline tree, but behaves in the following way:

- It does not open when you attempt to do so with a visibility cycling command (see Section 2.3 [Visibility cycling], page 6). You can force cycling archived subtrees with *C-TAB*, or by setting the option `org-cycle-open-archived-trees`. Also normal outline commands like `show-all` will open archived subtrees.
- During sparse tree construction (see Section 2.6 [Sparse trees], page 11), matches in archived subtrees are not exposed, unless you configure the option `org-sparse-tree-open-archived-trees`.
- During agenda view construction (see Chapter 10 [Agenda views], page 102), the content of archived trees is ignored unless you configure the option `org-agenda-skip-archived-trees`, in which case these trees will always be included. In the agenda you can press *v* `a` to get archives temporarily included.
- Archived trees are not exported (see Chapter 12 [Exporting], page 137), only the headline is. Configure the details using the variable `org-export-with-archived-trees`.
- Archived trees are excluded from column view unless the variable `org-columns-skip-archived-trees` is configured to `nil`.

The following commands help manage the ARCHIVE tag:

C-c C-x `a` — `org-toggle-archive-tag`
: Toggle the ARCHIVE tag for the current headline. When the tag is set, the headline changes to a shadowed face, and the subtree below it is hidden.

C-u C-c C-x `a`
: Check if any direct children of the current headline should be archived. To do this, each subtree is checked for open TODO entries. If none are found, the command offers to set the ARCHIVE tag for the child. If the cursor is *not* on a headline when this command is invoked, the level 1 trees will be checked.

C-TAB — `org-force-cycle-archived`
: Cycle a tree even if it is tagged with ARCHIVE.

C-c C-x A — `org-archive-to-archive-sibling`
: Move the current entry to the *Archive Sibling*. This is a sibling of the entry with the heading '`Archive`' and the tag '`ARCHIVE`'. The entry becomes a child of that sibling and in this way retains a lot of its original context, including inherited tags and approximate position in the outline.

10 Agenda views

Due to the way Org works, TODO items, time-stamped items, and tagged headlines can be scattered throughout a file or even a number of files. To get an overview of open action items, or of events that are important for a particular date, this information must be collected, sorted and displayed in an organized way.

Org can select items based on various criteria and display them in a separate buffer. Six different view types are provided:

- an *agenda* that is like a calendar and shows information for specific dates,
- a *TODO list* that covers all unfinished action items,
- a *match view*, showings headlines based on the tags, properties, and TODO state associated with them,
- a *text search view* that shows all entries from multiple files that contain specified keywords,
- a *stuck projects view* showing projects that currently don't move along, and
- *custom views* that are special searches and combinations of different views.

The extracted information is displayed in a special *agenda buffer*. This buffer is read-only, but provides commands to visit the corresponding locations in the original Org files, and even to edit these files remotely.

By default, the report ignores commented (see Section 12.6 [Comment lines], page 144) and archived (see Section 9.6.2 [Internal archiving], page 101) entries. You can override this by setting `org-agenda-skip-comment-trees` and `org-agenda-skip-archived-trees` to `nil`.

Two variables control how the agenda buffer is displayed and whether the window configuration is restored when the agenda exits: `org-agenda-window-setup` and `org-agenda-restore-windows-after-quit`.

10.1 Agenda files

The information to be shown is normally collected from all *agenda files*, the files listed in the variable `org-agenda-files`[1]. If a directory is part of this list, all files with the extension `.org` in this directory will be part of the list.

Thus, even if you only work with a single Org file, that file should be put into the list[2]. You can customize `org-agenda-files`, but the easiest way to maintain it is through the following commands

C-c [`org-agenda-file-to-front`

Add current file to the list of agenda files. The file is added to the front of the list. If it was already in the list, it is moved to the front. With a prefix argument, file is added/moved to the end.

---

[1] If the value of that variable is not a list, but a single file name, then the list of agenda files will be maintained in that external file.

[2] When using the dispatcher, pressing `<` before selecting a command will actually limit the command to the current file, and ignore `org-agenda-files` until the next dispatcher command.

`C-c ]` — `org-remove-file`
: Remove current file from the list of agenda files.

`C-'` — `org-cycle-agenda-files`
`C-,`
: Cycle through agenda file list, visiting one file after the other.

`M-x org-iswitchb RET`
: Command to use an `iswitchb`-like interface to switch to and between Org buffers.

The Org menu contains the current list of files and can be used to visit any of them.

If you would like to focus the agenda temporarily on a file not in this list, or on just one file in the list, or even on only a subtree in a file, then this can be done in different ways. For a single agenda command, you may press `<` once or several times in the dispatcher (see Section 10.2 [Agenda dispatcher], page 103). To restrict the agenda scope for an extended period, use the following commands:

`C-c C-x <` — `org-agenda-set-restriction-lock`
: Permanently restrict the agenda to the current subtree. When with a prefix argument, or with the cursor before the first headline in a file, the agenda scope is set to the entire file. This restriction remains in effect until removed with `C-c C-x >`, or by typing either `<` or `>` in the agenda dispatcher. If there is a window displaying an agenda view, the new restriction takes effect immediately.

`C-c C-x >` — `org-agenda-remove-restriction-lock`
: Remove the permanent restriction created by `C-c C-x <`.

When working with `speedbar.el`, you can use the following commands in the Speedbar frame:

`<` in the speedbar frame — `org-speedbar-set-agenda-restriction`
: Permanently restrict the agenda to the item—either an Org file or a subtree in such a file—at the cursor in the Speedbar frame. If there is a window displaying an agenda view, the new restriction takes effect immediately.

`>` in the speedbar frame — `org-agenda-remove-restriction-lock`
: Lift the restriction.

10.2 The agenda dispatcher

The views are created through a dispatcher, which should be bound to a global key—for example `C-c a` (see Section 1.3 [Activation], page 3). In the following we will assume that `C-c a` is indeed how the dispatcher is accessed and list keyboard access to commands accordingly. After pressing `C-c a`, an additional letter is required to execute a command. The dispatcher offers the following default commands:

`a`
: Create the calendar-like agenda (see Section 10.3.1 [Weekly/daily agenda], page 104).

`t / T`
: Create a list of all TODO items (see Section 10.3.2 [Global TODO list], page 106).

`m / M`
: Create a list of headlines matching a TAGS expression (see Section 10.3.3 [Matching tags and properties], page 107).

`s` Create a list of entries selected by a boolean expression of keywords and/or regular expressions that must or must not occur in the entry.

`/` Search for a regular expression in all agenda files and additionally in the files listed in `org-agenda-text-search-extra-files`. This uses the Emacs command `multi-occur`. A prefix argument can be used to specify the number of context lines for each match, default is 1.

`# / !` Create a list of stuck projects (see Section 10.3.5 [Stuck projects], page 110).

`<` Restrict an agenda command to the current buffer[3]. After pressing `<`, you still need to press the character selecting the command.

`< <` If there is an active region, restrict the following agenda command to the region. Otherwise, restrict it to the current subtree[4]. After pressing `< <`, you still need to press the character selecting the command.

`*` Toggle sticky agenda views. By default, Org maintains only a single agenda buffer and rebuilds it each time you change the view, to make sure everything is always up to date. If you often switch between agenda views and the build time bothers you, you can turn on sticky agenda buffers or make this the default by customizing the variable `org-agenda-sticky`. With sticky agendas, the agenda dispatcher will not recreate agenda views from scratch, it will only switch to the selected one, and you need to update the agenda by hand with *r* or *g* when needed. You can toggle sticky agenda view any time with `org-toggle-sticky-agenda`.

You can also define custom commands that will be accessible through the dispatcher, just like the default commands. This includes the possibility to create extended agenda buffers that contain several blocks together, for example the weekly agenda, the global TODO list and a number of special tags matches. See Section 10.6 [Custom agenda views], page 123.

10.3 The built-in agenda views

In this section we describe the built-in views.

10.3.1 The weekly/daily agenda

The purpose of the weekly/daily *agenda* is to act like a page of a paper agenda, showing all the tasks for the current week or day.

`C-c a a` `org-agenda-list`
Compile an agenda for the current week from a list of Org files. The agenda shows the entries for each day. With a numeric prefix[5] (like `C-u 2 1 C-c a a`) you may set the number of days to be displayed.

The default number of days displayed in the agenda is set by the variable `org-agenda-span` (or the obsolete `org-agenda-ndays`). This variable can be set to any number of

[3] For backward compatibility, you can also press `1` to restrict to the current buffer.

[4] For backward compatibility, you can also press `0` to restrict to the current region/subtree.

[5] For backward compatibility, the universal prefix `C-u` causes all TODO entries to be listed before the agenda. This feature is deprecated, use the dedicated TODO list, or a block agenda instead (see Section 10.6.2 [Block agenda], page 125).

days you want to see by default in the agenda, or to a span name, such as `day`, `week`, `month` or `year`. For weekly agendas, the default is to start on the previous monday (see `org-agenda-start-on-weekday`). You can also set the start date using a date shift: `(setq org-agenda-start-day "+10d")` will start the agenda ten days from today in the future.

Remote editing from the agenda buffer means, for example, that you can change the dates of deadlines and appointments from the agenda buffer. The commands available in the Agenda buffer are listed in Section 10.5 [Agenda commands], page 115.

Calendar/Diary integration

Emacs contains the calendar and diary by Edward M. Reingold. The calendar displays a three-month calendar with holidays from different countries and cultures. The diary allows you to keep track of anniversaries, lunar phases, sunrise/set, recurrent appointments (weekly, monthly) and more. In this way, it is quite complementary to Org. It can be very useful to combine output from Org with the diary.

In order to include entries from the Emacs diary into Org mode's agenda, you only need to customize the variable

```
(setq org-agenda-include-diary t)
```

After that, everything will happen automatically. All diary entries including holidays, anniversaries, etc., will be included in the agenda buffer created by Org mode. `SPC`, `TAB`, and `RET` can be used from the agenda buffer to jump to the diary file in order to edit existing diary entries. The *i* command to insert new entries for the current date works in the agenda buffer, as well as the commands *S*, *M*, and *C* to display Sunrise/Sunset times, show lunar phases and to convert to other calendars, respectively. `c` can be used to switch back and forth between calendar and agenda.

If you are using the diary only for sexp entries and holidays, it is faster to not use the above setting, but instead to copy or even move the entries into an Org file. Org mode evaluates diary-style sexp entries, and does it faster because there is no overhead for first creating the diary display. Note that the sexp entries must start at the left margin, no whitespace is allowed before them. For example, the following segment of an Org file will be processed and entries will be made in the agenda:

```
* Holidays
  :PROPERTIES:
  :CATEGORY: Holiday
  :END:
%%(org-calendar-holiday)   ; special function for holiday names

* Birthdays
  :PROPERTIES:
  :CATEGORY: Ann
  :END:
%%(org-anniversary 1956  5 14)[6] Arthur Dent is %d years old
%%(org-anniversary 1869 10  2) Mahatma Gandhi would be %d years old
```

[6] `org-anniversary` is just like `diary-anniversary`, but the argument order is always according to ISO and therefore independent of the value of `calendar-date-style`.

Anniversaries from BBDB

If you are using the Big Brothers Database to store your contacts, you will very likely prefer to store anniversaries in BBDB rather than in a separate Org or diary file. Org supports this and will show BBDB anniversaries as part of the agenda. All you need to do is to add the following to one of your agenda files:

```
* Anniversaries
  :PROPERTIES:
  :CATEGORY: Anniv
  :END:
%%(org-bbdb-anniversaries)
```

You can then go ahead and define anniversaries for a BBDB record. Basically, you need to press *C-o anniversary RET* with the cursor in a BBDB record and then add the date in the format `YYYY-MM-DD` or `MM-DD`, followed by a space and the class of the anniversary ('`birthday`' or '`wedding`', or a format string). If you omit the class, it will default to '`birthday`'. Here are a few examples, the header for the file `org-bbdb.el` contains more detailed information.

```
1973-06-22
06-22
1955-08-02 wedding
2008-04-14 %s released version 6.01 of org mode, %d years ago
```

After a change to BBDB, or for the first agenda display during an Emacs session, the agenda display will suffer a short delay as Org updates its hash with anniversaries. However, from then on things will be very fast—much faster in fact than a long list of '`%%(diary-anniversary)`' entries in an Org or Diary file.

If you would like to see upcoming anniversaries with a bit of forewarning, you can use the following instead:

```
* Anniversaries
  :PROPERTIES:
  :CATEGORY: Anniv
  :END:
%%(org-bbdb-anniversaries-future 3)
```

That will give you three days' warning: on the anniversary date itself and the two days prior. The argument is optional: if omitted, it defaults to 7.

Appointment reminders

Org can interact with Emacs appointments notification facility. To add the appointments of your agenda files, use the command `org-agenda-to-appt`. This command lets you filter through the list of your appointments and add only those belonging to a specific category or matching a regular expression. It also reads a `APPT_WARNTIME` property which will then override the value of `appt-message-warning-time` for this appointment. See the docstring for details.

10.3.2 The global TODO list

The global TODO list contains all unfinished TODO items formatted and collected into a single place.

`C-c a t` **org-todo-list**

Show the global TODO list. This collects the TODO items from all agenda files (see Chapter 10 [Agenda views], page 102) into a single buffer. By default, this lists items with a state the is not a DONE state. The buffer is in `agenda-mode`, so there are commands to examine and manipulate the TODO entries directly from that buffer (see Section 10.5 [Agenda commands], page 115).

`C-c a T` **org-todo-list**

Like the above, but allows selection of a specific TODO keyword. You can also do this by specifying a prefix argument to `C-c a t`. You are prompted for a keyword, and you may also specify several keywords by separating them with '`|`' as the boolean OR operator. With a numeric prefix, the Nth keyword in `org-todo-keywords` is selected. The `r` key in the agenda buffer regenerates it, and you can give a prefix argument to this command to change the selected TODO keyword, for example `3 r`. If you often need a search for a specific keyword, define a custom command for it (see Section 10.2 [Agenda dispatcher], page 103).

Matching specific TODO keywords can also be done as part of a tags search (see Section 6.4 [Tag searches], page 63).

Remote editing of TODO items means that you can change the state of a TODO entry with a single key press. The commands available in the TODO list are described in Section 10.5 [Agenda commands], page 115.

Normally the global TODO list simply shows all headlines with TODO keywords. This list can become very long. There are two ways to keep it more compact:

- Some people view a TODO item that has been *scheduled* for execution or have a *deadline* (see Section 8.1 [Timestamps], page 73) as no longer *open*. Configure the variables `org-agenda-todo-ignore-scheduled`, `org-agenda-todo-ignore-deadlines`, `org-agenda-todo-ignore-timestamp` and/or `org-agenda-todo-ignore-with-date` to exclude such items from the global TODO list.
- TODO items may have sublevels to break up the task into subtasks. In such cases it may be enough to list only the highest level TODO headline and omit the sublevels from the global list. Configure the variable `org-agenda-todo-list-sublevels` to get this behavior.

10.3.3 Matching tags and properties

If headlines in the agenda files are marked with *tags* (see Chapter 6 [Tags], page 59), or have properties (see Chapter 7 [Properties and columns], page 64), you can select headlines based on this metadata and collect them into an agenda buffer. The match syntax described here also applies when creating sparse trees with `C-c / m`.

`C-c a m` **org-tags-view**

Produce a list of all headlines that match a given set of tags. The command prompts for a selection criterion, which is a boolean logic expression with tags, like '`+work+urgent-withboss`' or '`work|home`' (see Chapter 6 [Tags], page 59). If you often need a specific search, define a custom command for it (see Section 10.2 [Agenda dispatcher], page 103).

`C-c a M` org-tags-view
: Like `C-c a m`, but only select headlines that are also TODO items in a not-DONE state and force checking subitems (see variable `org-tags-match-list-sublevels`). To exclude scheduled/deadline items, see the variable `org-agenda-tags-todo-honor-ignore-options`. Matching specific TODO keywords together with a tags match is also possible, see Section 6.4 [Tag searches], page 63.

The commands available in the tags list are described in Section 10.5 [Agenda commands], page 115.

Match syntax

A search string can use Boolean operators '`&`' for `AND` and '`|`' for `OR`. '`&`' binds more strongly than '`|`'. Parentheses are not implemented. Each element in the search is either a tag, a regular expression matching tags, or an expression like `PROPERTY OPERATOR VALUE` with a comparison operator, accessing a property value. Each element may be preceded by '`-`', to select against it, and '`+`' is syntactic sugar for positive selection. The `AND` operator '`&`' is optional when '`+`' or '`-`' is present. Here are some examples, using only tags.

'`work`' Select headlines tagged '`:work:`'.

'`work&boss`'
: Select headlines tagged '`:work:`' and '`:boss:`'.

'`+work-boss`'
: Select headlines tagged '`:work:`', but discard those also tagged '`:boss:`'.

'`work|laptop`'
: Selects lines tagged '`:work:`' or '`:laptop:`'.

'`work|laptop+night`'
: Like before, but require the '`:laptop:`' lines to be tagged also '`:night:`'.

Instead of a tag, you may also specify a regular expression enclosed in curly braces. For example, '`work+{^boss.*}`' matches headlines that contain the tag '`:work:`' and any tag *starting* with '`boss`'.

Group tags (see Section 6.3 [Tag hierarchy], page 62) are expanded as regular expressions. E.g., if '`:work:`' is a group tag for the group '`:work:lab:conf:`', then searching for '`work`' will search for '`{\(?:work\|lab\|conf\)}`' and searching for '`-work`' will search for all headlines but those with one of the tags in the group (i.e., '`-{\(?:work\|lab\|conf\)}`').

You may also test for properties (see Chapter 7 [Properties and columns], page 64) at the same time as matching tags. The properties may be real properties, or special properties that represent other metadata (see Section 7.2 [Special properties], page 66). For example, the "property" `TODO` represents the TODO keyword of the entry and the "property" `PRIORITY` represents the PRIORITY keyword of the entry.

In addition to the properties mentioned above, `LEVEL` represents the level of an entry. So a search '`+LEVEL=3+boss-TODO="DONE"`' lists all level three headlines that have the tag '`boss`' and are *not* marked with the TODO keyword DONE. In buffers with `org-odd-levels-only` set, '`LEVEL`' does not count the number of stars, but '`LEVEL=2`' will correspond to 3 stars etc.

Here are more examples:

'work+TODO="WAITING"'
: Select ':work:'-tagged TODO lines with the specific TODO keyword 'WAITING'.

'work+TODO="WAITING"|home+TODO="WAITING"'
: Waiting tasks both at work and at home.

When matching properties, a number of different operators can be used to test the value of a property. Here is a complex example:

```
+work-boss+PRIORITY="A"+Coffee="unlimited"+Effort<2         \
      +With={Sarah\|Denny}+SCHEDULED>="<2008-10-11>"
```

The type of comparison will depend on how the comparison value is written:

- If the comparison value is a plain number, a numerical comparison is done, and the allowed operators are '<', '=', '>', '<=', '>=', and '<>'.
- If the comparison value is enclosed in double-quotes, a string comparison is done, and the same operators are allowed.
- If the comparison value is enclosed in double-quotes *and* angular brackets (like 'DEADLINE<="<2008-12-24 18:30>"'), both values are assumed to be date/time specifications in the standard Org way, and the comparison will be done accordingly. Special values that will be recognized are "<now>" for now (including time), and "<today>", and "<tomorrow>" for these days at 00:00 hours, i.e., without a time specification. Also strings like "<+5d>" or "<-2m>" with units d, w, m, and y for day, week, month, and year, respectively, can be used.
- If the comparison value is enclosed in curly braces, a regexp match is performed, with '=' meaning that the regexp matches the property value, and '<>' meaning that it does not match.

So the search string in the example finds entries tagged ':work:' but not ':boss:', which also have a priority value 'A', a ':Coffee:' property with the value 'unlimited', an 'Effort' property that is numerically smaller than 2, a ':With:' property that is matched by the regular expression 'Sarah\|Denny', and that are scheduled on or after October 11, 2008.

You can configure Org mode to use property inheritance during a search, but beware that this can slow down searches considerably. See Section 7.4 [Property inheritance], page 67, for details.

For backward compatibility, and also for typing speed, there is also a different way to test TODO states in a search. For this, terminate the tags/property part of the search string (which may include several terms connected with '|') with a '/' and then specify a Boolean expression just for TODO keywords. The syntax is then similar to that for tags, but should be applied with care: for example, a positive selection on several TODO keywords cannot meaningfully be combined with boolean AND. However, *negative selection* combined with AND can be meaningful. To make sure that only lines are checked that actually have any TODO keyword (resulting in a speed-up), use *C-c* a *M*, or equivalently start the TODO part after the slash with '!'. Using *C-c* a *M* or '/!' will not match TODO keywords in a DONE state. Examples:

'work/WAITING'
: Same as 'work+TODO="WAITING"'

'work/!-WAITING-NEXT'
: Select ':work:'-tagged TODO lines that are neither 'WAITING' nor 'NEXT'

'work/!+WAITING|+NEXT'
: Select ':work:'-tagged TODO lines that are either 'WAITING' or 'NEXT'.

10.3.4 Search view

This agenda view is a general text search facility for Org mode entries. It is particularly useful to find notes.

C-c a s `org-search-view`
: This is a special search that lets you select entries by matching a substring or specific words using a boolean logic.

For example, the search string 'computer equipment' will find entries that contain 'computer equipment' as a substring. If the two words are separated by more space or a line break, the search will still match. Search view can also search for specific keywords in the entry, using Boolean logic. The search string '+computer +wifi -ethernet -{8\.11[bg]}' will search for note entries that contain the keywords `computer` and `wifi`, but not the keyword `ethernet`, and which are also not matched by the regular expression `8\.11[bg]`, meaning to exclude both 8.11b and 8.11g. The first '+' is necessary to turn on word search, other '+' characters are optional. For more details, see the docstring of the command `org-search-view`.

Note that in addition to the agenda files, this command will also search the files listed in `org-agenda-text-search-extra-files`.

10.3.5 Stuck projects

If you are following a system like David Allen's GTD to organize your work, one of the "duties" you have is a regular review to make sure that all projects move along. A *stuck* project is a project that has no defined next actions, so it will never show up in the TODO lists Org mode produces. During the review, you need to identify such projects and define next actions for them.

C-c a # `org-agenda-list-stuck-projects`
: List projects that are stuck.

C-c a ! Customize the variable `org-stuck-projects` to define what a stuck project is and how to find it.

You almost certainly will have to configure this view before it will work for you. The built-in default assumes that all your projects are level-2 headlines, and that a project is not stuck if it has at least one entry marked with a TODO keyword TODO or NEXT or NEXTACTION.

Let's assume that you, in your own way of using Org mode, identify projects with a tag PROJECT, and that you use a TODO keyword MAYBE to indicate a project that should not be considered yet. Let's further assume that the TODO keyword DONE marks finished projects, and that NEXT and TODO indicate next actions. The tag @SHOP indicates shopping and is a next action even without the NEXT tag. Finally, if the project contains the special word IGNORE anywhere, it should not be listed either. In this case you would start by identifying eligible projects with a tags/todo match[7] '+PROJECT/-MAYBE-DONE', and then check for TODO, NEXT, @SHOP, and IGNORE in the subtree to identify projects that are not stuck. The correct customization for this is

```
(setq org-stuck-projects
      '("+PROJECT/-MAYBE-DONE" ("NEXT" "TODO") ("@SHOP")
                               "\\<IGNORE\\>"))
```

Note that if a project is identified as non-stuck, the subtree of this entry will still be searched for stuck projects.

10.4 Presentation and sorting

Before displaying items in an agenda view, Org mode visually prepares the items and sorts them. Each item occupies a single line. The line starts with a *prefix* that contains the *category* (see Section 10.4.1 [Categories], page 111) of the item and other important information. You can customize in which column tags will be displayed through `org-agenda-tags-column`. You can also customize the prefix using the option `org-agenda-prefix-format`. This prefix is followed by a cleaned-up version of the outline headline associated with the item.

10.4.1 Categories

The category is a broad label assigned to each agenda item. By default, the category is simply derived from the file name, but you can also specify it with a special line in the buffer, like this:

```
#+CATEGORY: Thesis
```

If you would like to have a special CATEGORY for a single entry or a (sub)tree, give the entry a `:CATEGORY:` property with the special category you want to apply as the value.

The display in the agenda buffer looks best if the category is not longer than 10 characters.

You can set up icons for category by customizing the `org-agenda-category-icon-alist` variable.

10.4.2 Time-of-day specifications

Org mode checks each agenda item for a time-of-day specification. The time can be part of the timestamp that triggered inclusion into the agenda, for example as in '<2005-05-10 Tue 19:00>'. Time ranges can be specified with two timestamps, like '<2005-05-10 Tue 20:30>--<2005-05-10 Tue 22:15>'.

In the headline of the entry itself, a time(range) may also appear as plain text (like '12:45' or a '8:30-1pm'). If the agenda integrates the Emacs diary (see Section 10.3.1 [Weekly/daily agenda], page 104), time specifications in diary entries are recognized as well.

For agenda display, Org mode extracts the time and displays it in a standard 24 hour format as part of the prefix. The example times in the previous paragraphs would end up in the agenda like this:

---

[7] See Section 6.4 [Tag searches], page 63.

```
 8:30-13:00 Arthur Dent lies in front of the bulldozer
12:45...... Ford Prefect arrives and takes Arthur to the pub
19:00...... The Vogon reads his poem
20:30-22:15 Marvin escorts the Hitchhikers to the bridge
```

If the agenda is in single-day mode, or for the display of today, the timed entries are embedded in a time grid, like

```
 8:00...... ------------------
 8:30-13:00 Arthur Dent lies in front of the bulldozer
10:00...... ------------------
12:00...... ------------------
12:45...... Ford Prefect arrives and takes Arthur to the pub
14:00...... ------------------
16:00...... ------------------
18:00...... ------------------
19:00...... The Vogon reads his poem
20:00...... ------------------
20:30-22:15 Marvin escorts the Hitchhikers to the bridge
```

The time grid can be turned on and off with the variable `org-agenda-use-time-grid`, and can be configured with `org-agenda-time-grid`.

10.4.3 Sorting agenda items

Before being inserted into a view, the items are sorted. How this is done depends on the type of view.

- For the daily/weekly agenda, the items for each day are sorted. The default order is to first collect all items containing an explicit time-of-day specification. These entries will be shown at the beginning of the list, as a *schedule* for the day. After that, items remain grouped in categories, in the sequence given by `org-agenda-files`. Within each category, items are sorted by priority (see Section 5.4 [Priorities], page 55), which is composed of the base priority (2000 for priority '`A`', 1000 for '`B`', and 0 for '`C`'), plus additional increments for overdue scheduled or deadline items.
- For the TODO list, items remain in the order of categories, but within each category, sorting takes place according to priority (see Section 5.4 [Priorities], page 55). The priority used for sorting derives from the priority cookie, with additions depending on how close an item is to its due or scheduled date.
- For tags matches, items are not sorted at all, but just appear in the sequence in which they are found in the agenda files.

Sorting can be customized using the variable `org-agenda-sorting-strategy`, and may also include criteria based on the estimated effort of an entry (see Section 8.5 [Effort estimates], page 86).

10.4.4 Filtering/limiting agenda items

Agenda built-in or customized commands are statically defined. Agenda filters and limits provide two ways of dynamically narrowing down the list of agenda entries: *filters* and *limits*. Filters only act on the display of the items, while limits take effect before the list of agenda entries is built. Filters are more often used interactively, while limits are mostly useful when defined as local variables within custom agenda commands.

Filtering in the agenda

`/` `org-agenda-filter-by-tag`

Filter the agenda view with respect to a tag and/or effort estimates. The difference between this and a custom agenda command is that filtering is very fast, so that you can switch quickly between different filters without having to recreate the agenda.[8]

You will be prompted for a tag selection letter; `SPC` will mean any tag at all. Pressing `TAB` at that prompt will offer use completion to select a tag (including any tags that do not have a selection character). The command then hides all entries that do not contain or inherit this tag. When called with prefix arg, remove the entries that *do* have the tag. A second `/` at the prompt will turn off the filter and unhide any hidden entries. Pressing `+` or `-` switches between filtering and excluding the next tag.

Org also supports automatic, context-aware tag filtering. If the variable `org-agenda-auto-exclude-function` is set to a user-defined function, that function can decide which tags should be excluded from the agenda automatically. Once this is set, the `/` command then accepts *RET* as a sub-option key and runs the auto exclusion logic. For example, let's say you use a `Net` tag to identify tasks which need network access, an `Errand` tag for errands in town, and a `Call` tag for making phone calls. You could auto-exclude these tags based on the availability of the Internet, and outside of business hours, with something like this:

```
(defun org-my-auto-exclude-function (tag)
  (and (cond
        ((string= tag "Net")
         (/= 0 (call-process "/sbin/ping" nil nil nil
                             "-c1" "-q" "-t1" "mail.gnu.org")))
        ((or (string= tag "Errand") (string= tag "Call"))
         (let ((hour (nth 2 (decode-time))))
           (or (< hour 8) (> hour 21)))))
       (concat "-" tag)))

(setq org-agenda-auto-exclude-function 'org-my-auto-exclude-function)
```

`[ ] { }`

in *search view*

add new search words (`[` and `]`) or new regular expressions (`{` and `}`) to the query string. The opening bracket/brace will add a positive search term prefixed by '+', indicating that this search term *must* occur/match in the entry. The closing bracket/brace will add a negative search term which *must not* occur/match in the entry for it to be selected.

---

[8] Custom commands can preset a filter by binding the variable `org-agenda-tag-filter-preset` as an option. This filter will then be applied to the view and persist as a basic filter through refreshes and more secondary filtering. The filter is a global property of the entire agenda view—in a block agenda, you should only set this in the global options section, not in the section of an individual block.

< org-agenda-filter-by-category
: Filter the current agenda view with respect to the category of the item at point. Pressing < another time will remove this filter. When called with a prefix argument exclude the category of the item at point from the agenda.

You can add a filter preset in custom agenda commands through the option `org-agenda-category-filter-preset`. See Section 10.6.3 [Setting options], page 125.

^ org-agenda-filter-by-top-headline
: Filter the current agenda view and only display the siblings and the parent headline of the one at point.

= org-agenda-filter-by-regexp
: Filter the agenda view by a regular expression: only show agenda entries matching the regular expression the user entered. When called with a prefix argument, it will filter *out* entries matching the regexp. With two universal prefix arguments, it will remove all the regexp filters, which can be accumulated.

You can add a filter preset in custom agenda commands through the option `org-agenda-regexp-filter-preset`. See Section 10.6.3 [Setting options], page 125.

_ org-agenda-filter-by-effort
: Filter the agenda view with respect to effort estimates. You first need to set up allowed efforts globally, for example

```
(setq org-global-properties
    '(("Effort_ALL". "0 0:10 0:30 1:00 2:00 3:00 4:00")))
```

You can then filter for an effort by first typing an operator, one of <, >, and =, and then the one-digit index of an effort estimate in your array of allowed values, where *0* means the 10th value. The filter will then restrict to entries with effort smaller-or-equal, equal, or larger-or-equal than the selected value. For application of the operator, entries without a defined effort will be treated according to the value of `org-sort-agenda-noeffort-is-high`.

When called with a prefix argument, it will remove entries matching the condition. With two universal prefix arguments, it will clear effort filters, which can be accumulated.

You can add a filter preset in custom agenda commands through the option `org-agenda-effort-filter-preset`. See Section 10.6.3 [Setting options], page 125.

| org-agenda-filter-remove-all
: Remove all filters in the current agenda view.

Setting limits for the agenda

Here is a list of options that you can set, either globally, or locally in your custom agenda views (see Section 10.6 [Custom agenda views], page 123).

`org-agenda-max-entries`
: Limit the number of entries.

`org-agenda-max-effort`
Limit the duration of accumulated efforts (as minutes).

`org-agenda-max-todos`
Limit the number of entries with TODO keywords.

`org-agenda-max-tags`
Limit the number of tagged entries.

When set to a positive integer, each option will exclude entries from other categories: for example, `(setq org-agenda-max-effort 100)` will limit the agenda to 100 minutes of effort and exclude any entry that has no effort property. If you want to include entries with no effort property, use a negative value for `org-agenda-max-effort`.

One useful setup is to use `org-agenda-max-entries` locally in a custom command. For example, this custom command will display the next five entries with a `NEXT` TODO keyword.

```
(setq org-agenda-custom-commands
      '(("n" todo "NEXT"
         ((org-agenda-max-entries 5)))))
```

Once you mark one of these five entry as `DONE`, rebuilding the agenda will again the next five entries again, including the first entry that was excluded so far.

You can also dynamically set temporary limits, which will be lost when rebuilding the agenda:

~ `org-agenda-limit-interactively`
This prompts for the type of limit to apply and its value.

10.5 Commands in the agenda buffer

Entries in the agenda buffer are linked back to the Org file or diary file where they originate. You are not allowed to edit the agenda buffer itself, but commands are provided to show and jump to the original entry location, and to edit the Org files "remotely" from the agenda buffer. In this way, all information is stored only once, removing the risk that your agenda and note files may diverge.

Some commands can be executed with mouse clicks on agenda lines. For the other commands, the cursor needs to be in the desired line.

Motion

n `org-agenda-next-line`
Next line (same as `down` and *C-n*).

p `org-agenda-previous-line`
Previous line (same as `up` and *C-p*).

N `org-agenda-next-item`
Next item: same as next line, but only consider items.

P `org-agenda-previous-item`
Previous item: same as previous line, but only consider items.

View/Go to Org file

SPC or *mouse-3* `org-agenda-show-and-scroll-up`
: Display the original location of the item in another window. With prefix arg, make sure that drawers stay folded.

L `org-agenda-recenter`
: Display original location and recenter that window.

TAB or *mouse-2* `org-agenda-goto`
: Go to the original location of the item in another window.

RET `org-agenda-switch-to`
: Go to the original location of the item and delete other windows.

F `org-agenda-follow-mode`
: Toggle Follow mode. In Follow mode, as you move the cursor through the agenda buffer, the other window always shows the corresponding location in the Org file. The initial setting for this mode in new agenda buffers can be set with the variable `org-agenda-start-with-follow-mode`.

C-c C-x b `org-agenda-tree-to-indirect-buffer`
: Display the entire subtree of the current item in an indirect buffer. With a numeric prefix argument N, go up to level N and then take that tree. If N is negative, go up that many levels. With a *C-u* prefix, do not remove the previously used indirect buffer.

C-c C-o `org-agenda-open-link`
: Follow a link in the entry. This will offer a selection of any links in the text belonging to the referenced Org node. If there is only one link, it will be followed without a selection prompt.

Change display

A Interactively select another agenda view and append it to the current view.

o Delete other windows.

v d or short *d* `org-agenda-day-view`
v w or short *w* `org-agenda-week-view`
v t `org-agenda-fortnight-view`
v m `org-agenda-month-view`
v y `org-agenda-year-view`
v SPC `org-agenda-reset-view`
: Switch to day/week/month/year view. When switching to day or week view, this setting becomes the default for subsequent agenda refreshes. Since month and year views are slow to create, they do not become the default. A numeric prefix argument may be used to jump directly to a specific day of the year, ISO week, month, or year, respectively. For example, *32 d* jumps to February 1st, *9 w* to ISO week number 9. When setting day, week, or month view, a year may be encoded in the prefix argument as well. For example, *200712 w* will jump to week 12 in 2007. If such a year specification has only one or two digits, it will be mapped to the interval 1938–2037. *v SPC* will reset to what is set in `org-agenda-span`.

f `org-agenda-later`
: Go forward in time to display the following `org-agenda-current-span` days. For example, if the display covers a week, switch to the following week. With prefix arg, go forward that many times `org-agenda-current-span` days.

b `org-agenda-earlier`
: Go backward in time to display earlier dates.

. `org-agenda-goto-today`
: Go to today.

j `org-agenda-goto-date`
: Prompt for a date and go there.

J `org-agenda-clock-goto`
: Go to the currently clocked-in task *in the agenda buffer.*

D `org-agenda-toggle-diary`
: Toggle the inclusion of diary entries. See Section 10.3.1 [Weekly/daily agenda], page 104.

v l or short *l* `org-agenda-log-mode`
: Toggle Logbook mode. In Logbook mode, entries that were marked DONE while logging was on (variable `org-log-done`) are shown in the agenda, as are entries that have been clocked on that day. You can configure the entry types that should be included in log mode using the variable `org-agenda-log-mode-items`. When called with a *C-u* prefix, show all possible logbook entries, including state changes. When called with two prefix arguments *C-u C-u*, show only logging information, nothing else. *v L* is equivalent to *C-u v l*.

v [or short *[* `org-agenda-manipulate-query-add`
: Include inactive timestamps into the current view. Only for weekly/daily agenda.

v a `org-agenda-archives-mode`
v A `org-agenda-archives-mode 'files`
: Toggle Archives mode. In Archives mode, trees that are marked `ARCHIVED` are also scanned when producing the agenda. When you use the capital *A*, even all archive files are included. To exit archives mode, press *v a* again.

v R or short *R* `org-agenda-clockreport-mode`
: Toggle Clockreport mode. In Clockreport mode, the daily/weekly agenda will always show a table with the clocked times for the time span and file scope covered by the current agenda view. The initial setting for this mode in new agenda buffers can be set with the variable `org-agenda-start-with-clockreport-mode`. By using a prefix argument when toggling this mode (i.e., *C-u R*), the clock table will not show contributions from entries that are hidden by agenda filtering[9]. See also the variable `org-clock-report-include-clocking-task`.

v c
: Show overlapping clock entries, clocking gaps, and other clocking problems in the current agenda range. You can then visit clocking lines and fix them manually. See the variable `org-agenda-clock-consistency-checks` for information

[9] Only tags filtering will be respected here, effort filtering is ignored.

on how to customize the definition of what constituted a clocking problem. To return to normal agenda display, press `l` to exit Logbook mode.

v E or short *E* `org-agenda-entry-text-mode`
: Toggle entry text mode. In entry text mode, a number of lines from the Org outline node referenced by an agenda line will be displayed below the line. The maximum number of lines is given by the variable `org-agenda-entry-text-maxlines`. Calling this command with a numeric prefix argument will temporarily modify that number to the prefix value.

G `org-agenda-toggle-time-grid`
: Toggle the time grid on and off. See also the variables `org-agenda-use-time-grid` and `org-agenda-time-grid`.

r `org-agenda-redo`
: Recreate the agenda buffer, for example to reflect the changes after modification of the timestamps of items with *S-left* and *S-right*. When the buffer is the global TODO list, a prefix argument is interpreted to create a selective list for a specific TODO keyword.

g `org-agenda-redo`
: Same as *r*.

C-x C-s or short *s* `org-save-all-org-buffers`
: Save all Org buffers in the current Emacs session, and also the locations of IDs.

C-c C-x C-c `org-agenda-columns`
: Invoke column view (see Section 7.5 [Column view], page 67) in the agenda buffer. The column view format is taken from the entry at point, or (if there is no entry at point), from the first entry in the agenda view. So whatever the format for that entry would be in the original buffer (taken from a property, from a `#+COLUMNS` line, or from the default variable `org-columns-default-format`), will be used in the agenda.

C-c C-x > `org-agenda-remove-restriction-lock`
: Remove the restriction lock on the agenda, if it is currently restricted to a file or subtree (see Section 10.1 [Agenda files], page 102).

Secondary filtering and query editing
: For a detailed description of these commands, see Section 10.4.4 [Filtering/limiting agenda items], page 112.

/ `org-agenda-filter-by-tag`
: Filter the agenda view with respect to a tag and/or effort estimates.

< `org-agenda-filter-by-category`
: Filter the current agenda view with respect to the category of the item at point.

^ `org-agenda-filter-by-top-headline`
: Filter the current agenda view and only display the siblings and the parent headline of the one at point.

= org-agenda-filter-by-regexp
: Filter the agenda view by a regular expression.

_ org-agenda-filter-by-effort
: Filter the agenda view with respect to effort estimates.

| org-agenda-filter-remove-all
: Remove all filters in the current agenda view.

Remote editing

0--9 Digit argument.

C-_ org-agenda-undo
: Undo a change due to a remote editing command. The change is undone both in the agenda buffer and in the remote buffer.

t org-agenda-todo
: Change the TODO state of the item, both in the agenda and in the original org file.

C-S-right org-agenda-todo-nextset
C-S-left org-agenda-todo-previousset
: Switch to the next/previous set of TODO keywords.

C-k org-agenda-kill
: Delete the current agenda item along with the entire subtree belonging to it in the original Org file. If the text to be deleted remotely is longer than one line, the kill needs to be confirmed by the user. See variable `org-agenda-confirm-kill`.

C-c C-w org-agenda-refile
: Refile the entry at point.

C-c C-x C-a or short `a` org-agenda-archive-default-with-confirmation
: Archive the subtree corresponding to the entry at point using the default archiving command set in `org-archive-default-command`. When using the a key, confirmation will be required.

C-c C-x a org-agenda-toggle-archive-tag
: Toggle the ARCHIVE tag for the current headline.

C-c C-x A org-agenda-archive-to-archive-sibling
: Move the subtree corresponding to the current entry to its *archive sibling*.

C-c C-x C-s or short *$* org-agenda-archive
: Archive the subtree corresponding to the current headline. This means the entry will be moved to the configured archive location, most likely a different file.

T org-agenda-show-tags
: Show all tags associated with the current item. This is useful if you have turned off `org-agenda-show-inherited-tags`, but still want to see all tags of a headline occasionally.

: `org-agenda-set-tags`
: Set tags for the current headline. If there is an active region in the agenda, change a tag for all headings in the region.

, Set the priority for the current item (`org-agenda-priority`). Org mode prompts for the priority character. If you reply with `SPC`, the priority cookie is removed from the entry.

P `org-agenda-show-priority`
: Display weighted priority of current item.

`+` or *S-up* `org-agenda-priority-up`
: Increase the priority of the current item. The priority is changed in the original buffer, but the agenda is not resorted. Use the *r* key for this.

`-` or *S-down* `org-agenda-priority-down`
: Decrease the priority of the current item.

z or *C-c C-z* `org-agenda-add-note`
: Add a note to the entry. This note will be recorded, and then filed to the same location where state change notes are put. Depending on `org-log-into-drawer`, this may be inside a drawer.

C-c C-a `org-attach`
: Dispatcher for all command related to attachments.

C-c C-s `org-agenda-schedule`
: Schedule this item. With prefix arg remove the scheduling timestamp

C-c C-d `org-agenda-deadline`
: Set a deadline for this item. With prefix arg remove the deadline.

S-right `org-agenda-do-date-later`
: Change the timestamp associated with the current line by one day into the future. If the date is in the past, the first call to this command will move it to today.
With a numeric prefix argument, change it by that many days. For example, *3 6 5 S-right* will change it by a year. With a *C-u* prefix, change the time by one hour. If you immediately repeat the command, it will continue to change hours even without the prefix arg. With a double *C-u C-u* prefix, do the same for changing minutes.
The stamp is changed in the original Org file, but the change is not directly reflected in the agenda buffer. Use *r* or *g* to update the buffer.

S-left `org-agenda-do-date-earlier`
: Change the timestamp associated with the current line by one day into the past.

`>` `org-agenda-date-prompt`
: Change the timestamp associated with the current line. The key `>` has been chosen, because it is the same as *S-.* on my keyboard.

I `org-agenda-clock-in`
: Start the clock on the current item. If a clock is running already, it is stopped first.

O **org-agenda-clock-out**
: Stop the previously started clock.

X **org-agenda-clock-cancel**
: Cancel the currently running clock.

J **org-agenda-clock-goto**
: Jump to the running clock in another window.

k **org-agenda-capture**
: Like `org-capture`, but use the date at point as the default date for the capture template. See `org-capture-use-agenda-date` to make this the default behavior of `org-capture`.

Dragging agenda lines forward/backward

M-<up> **org-agenda-drag-line-backward**
: Drag the line at point backward one line[10]. With a numeric prefix argument, drag backward by that many lines.

M-<down> **org-agenda-drag-line-forward**
: Drag the line at point forward one line. With a numeric prefix argument, drag forward by that many lines.

Bulk remote editing selected entries

m **org-agenda-bulk-mark**
: Mark the entry at point for bulk action. With numeric prefix argument, mark that many successive entries.

*** **org-agenda-bulk-mark-all**
: Mark all visible agenda entries for bulk action.

u **org-agenda-bulk-unmark**
: Unmark entry at point for bulk action.

U **org-agenda-bulk-remove-all-marks**
: Unmark all marked entries for bulk action.

M-m **org-agenda-bulk-toggle**
: Toggle mark of the entry at point for bulk action.

*M-** **org-agenda-bulk-toggle-all**
: Toggle marks of all visible entries for bulk action.

% **org-agenda-bulk-mark-regexp**
: Mark entries matching a regular expression for bulk action.

B **org-agenda-bulk-action**
: Bulk action: act on all marked entries in the agenda. This will prompt for another key to select the action to be applied. The prefix arg to *B* will be passed through to the *s* and *d* commands, to bulk-remove these special timestamps. By default, marks are removed after the bulk. If you want them to persist, set `org-agenda-persistent-marks` to `t` or hit *p* at the prompt.

---

[10] Moving agenda lines does not persist after an agenda refresh and does not modify the contributing `.org` files

*   Toggle persistent marks.

$ Archive all selected entries.

A   Archive entries by moving them to their respective archive siblings.

t   Change TODO state. This prompts for a single TODO keyword and changes the state of all selected entries, bypassing blocking and suppressing logging notes (but not timestamps).

+   Add a tag to all selected entries.

-   Remove a tag from all selected entries.

s   Schedule all items to a new date. To shift existing schedule dates by a fixed number of days, use something starting with double plus at the prompt, for example '++8d' or '++2w'.

d   Set deadline to a specific date.

r   Prompt for a single refile target and move all entries. The entries will no longer be in the agenda; refresh (*g*) to bring them back.

S   Reschedule randomly into the coming N days. N will be prompted for. With prefix arg (*C-u B S*), scatter only across weekdays.

f   Apply a function[11] to marked entries. For example, the function below sets the CATEGORY property of the entries to web.

```
(defun set-category ()
  (interactive "P")
  (let* ((marker (or (org-get-at-bol 'org-hd-marker)
                     (org-agenda-error)))
         (buffer (marker-buffer marker)))
    (with-current-buffer buffer
      (save-excursion
        (save-restriction
          (widen)
          (goto-char marker)
          (org-back-to-heading t)
          (org-set-property "CATEGORY" "web"))))))
```

Calendar commands

c   `org-agenda-goto-calendar`

Open the Emacs calendar and move to the date at the agenda cursor.

c   `org-calendar-goto-agenda`

When in the calendar, compute and show the Org mode agenda for the date at the cursor.

i   `org-agenda-diary-entry`

Insert a new entry into the diary, using the date at the cursor and (for block entries) the date at the mark. This will add to the Emacs diary file[12], in a way

[11] You can also create persistent custom functions through `org-agenda-bulk-custom-functions`.

[12] This file is parsed for the agenda when `org-agenda-include-diary` is set.

similar to the *i* command in the calendar. The diary file will pop up in another window, where you can add the entry.

If you configure `org-agenda-diary-file` to point to an Org mode file, Org will create entries (in Org mode syntax) in that file instead. Most entries will be stored in a date-based outline tree that will later make it easy to archive appointments from previous months/years. The tree will be built under an entry with a `DATE_TREE` property, or else with years as top-level entries. Emacs will prompt you for the entry text—if you specify it, the entry will be created in `org-agenda-diary-file` without further interaction. If you directly press `RET` at the prompt without typing text, the target file will be shown in another window for you to finish the entry there. See also the *k r* command.

M `org-agenda-phases-of-moon`
Show the phases of the moon for the three months around current date.

S `org-agenda-sunrise-sunset`
Show sunrise and sunset times. The geographical location must be set with calendar variables, see the documentation for the Emacs calendar.

C `org-agenda-convert-date`
Convert the date at cursor into many other cultural and historic calendars.

H `org-agenda-holidays`
Show holidays for three months around the cursor date.

M-x org-icalendar-combine-agenda-files RET
Export a single iCalendar file containing entries from all agenda files. This is a globally available command, and also available in the agenda menu.

Exporting to a file

C-x C-w `org-agenda-write`
Write the agenda view to a file. Depending on the extension of the selected file name, the view will be exported as HTML (`.html` or `.htm`), Postscript (`.ps`), PDF (`.pdf`), Org (`.org`) and plain text (any other extension). When exporting to Org, only the body of original headlines are exported, not subtrees or inherited tags. When called with a *C-u* prefix argument, immediately open the newly created file. Use the variable `org-agenda-exporter-settings` to set options for `ps-print` and for `htmlize` to be used during export.

Quit and Exit

q `org-agenda-quit`
Quit agenda, remove the agenda buffer.

x `org-agenda-exit`
Exit agenda, remove the agenda buffer and all buffers loaded by Emacs for the compilation of the agenda. Buffers created by the user to visit Org files will not be removed.

10.6 Custom agenda views

Custom agenda commands serve two purposes: to store and quickly access frequently used TODO and tags searches, and to create special composite agenda buffers. Custom agenda commands will be accessible through the dispatcher (see Section 10.2 [Agenda dispatcher], page 103), just like the default commands.

10.6.1 Storing searches

The first application of custom searches is the definition of keyboard shortcuts for frequently used searches, either creating an agenda buffer, or a sparse tree (the latter covering of course only the current buffer).

Custom commands are configured in the variable `org-agenda-custom-commands`. You can customize this variable, for example by pressing *C-c a C*. You can also directly set it with Emacs Lisp in the Emacs init file. The following example contains all valid agenda views:

```
(setq org-agenda-custom-commands
      '(("x" agenda)
        ("y" agenda*)
        ("w" todo "WAITING")
        ("W" todo-tree "WAITING")
        ("u" tags "+boss-urgent")
        ("v" tags-todo "+boss-urgent")
        ("U" tags-tree "+boss-urgent")
        ("f" occur-tree "\\<FIXME\\>")
        ("h" . "HOME+Name tags searches") ; description for "h" prefix
        ("hl" tags "+home+Lisa")
        ("hp" tags "+home+Peter")
        ("hk" tags "+home+Kim")))
```

The initial string in each entry defines the keys you have to press after the dispatcher command *C-c a* in order to access the command. Usually this will be just a single character, but if you have many similar commands, you can also define two-letter combinations where the first character is the same in several combinations and serves as a prefix key[13]. The second parameter is the search type, followed by the string or regular expression to be used for the matching. The example above will therefore define:

C-c a x as a global search for agenda entries planned[14] this week/day.

C-c a y as a global search for agenda entries planned this week/day, but only those with an hour specification like `[h]h:mm`—think of them as appointments.

C-c a w as a global search for TODO entries with '`WAITING`' as the TODO keyword

C-c a W as the same search, but only in the current buffer and displaying the results as a sparse tree

C-c a u as a global tags search for headlines marked '`:boss:`' but not '`:urgent:`'

C-c a v as the same search as *C-c a u*, but limiting the search to headlines that are also TODO items

---

[13] You can provide a description for a prefix key by inserting a cons cell with the prefix and the description.

[14] *Planned* means here that these entries have some planning information attached to them, like a time-stamp, a scheduled or a deadline string. See `org-agenda-entry-types` on how to set what planning information will be taken into account.

C-c a U as the same search as *C-c a u*, but only in the current buffer and displaying the result as a sparse tree

C-c a f to create a sparse tree (again: current buffer only) with all entries containing the word 'FIXME'

C-c a h as a prefix command for a HOME tags search where you have to press an additional key (*l*, *p* or *k*) to select a name (Lisa, Peter, or Kim) as additional tag to match.

Note that the `*-tree` agenda views need to be called from an Org buffer as they operate on the current buffer only.

10.6.2 Block agenda

Another possibility is the construction of agenda views that comprise the results of *several* commands, each of which creates a block in the agenda buffer. The available commands include `agenda` for the daily or weekly agenda (as created with *C-c a a*), `alltodo` for the global TODO list (as constructed with *C-c a t*), and the matching commands discussed above: `todo`, `tags`, and `tags-todo`. Here are two examples:

```
(setq org-agenda-custom-commands
      '(("h" "Agenda and Home-related tasks"
         ((agenda "")
          (tags-todo "home")
          (tags "garden")))
        ("o" "Agenda and Office-related tasks"
         ((agenda "")
          (tags-todo "work")
          (tags "office")))))
```

This will define *C-c a h* to create a multi-block view for stuff you need to attend to at home. The resulting agenda buffer will contain your agenda for the current week, all TODO items that carry the tag 'home', and also all lines tagged with 'garden'. Finally the command *C-c a o* provides a similar view for office tasks.

10.6.3 Setting options for custom commands

Org mode contains a number of variables regulating agenda construction and display. The global variables define the behavior for all agenda commands, including the custom commands. However, if you want to change some settings just for a single custom view, you can do so. Setting options requires inserting a list of variable names and values at the right spot in `org-agenda-custom-commands`. For example:

```
(setq org-agenda-custom-commands
      '(("w" todo "WAITING"
         ((org-agenda-sorting-strategy '(priority-down))
          (org-agenda-prefix-format "  Mixed: ")))
        ("U" tags-tree "+boss-urgent"
         ((org-show-context-detail 'minimal)))
        ("N" search ""
         ((org-agenda-files '("~org/notes.org"))
          (org-agenda-text-search-extra-files nil)))))
```

Now the *C-c a w* command will sort the collected entries only by priority, and the prefix format is modified to just say '`Mixed:` ' instead of giving the category of the entry. The sparse tags tree of *C-c a U* will now turn out ultra-compact, because neither the headline hierarchy above the match, nor the headline following the match will be shown. The command *C-c a N* will do a text search limited to only a single file.

For command sets creating a block agenda, `org-agenda-custom-commands` has two separate spots for setting options. You can add options that should be valid for just a single command in the set, and options that should be valid for all commands in the set. The former are just added to the command entry; the latter must come after the list of command entries. Going back to the block agenda example (see Section 10.6.2 [Block agenda], page 125), let's change the sorting strategy for the *C-c a h* commands to `priority-down`, but let's sort the results for GARDEN tags query in the opposite order, `priority-up`. This would look like this:

```
(setq org-agenda-custom-commands
      '(("h" "Agenda and Home-related tasks"
         ((agenda)
          (tags-todo "home")
          (tags "garden"
                ((org-agenda-sorting-strategy '(priority-up)))))
         ((org-agenda-sorting-strategy '(priority-down))))
        ("o" "Agenda and Office-related tasks"
         ((agenda)
          (tags-todo "work")
          (tags "office")))))
```

As you see, the values and parentheses setting is a little complex. When in doubt, use the customize interface to set this variable—it fully supports its structure. Just one caveat: when setting options in this interface, the *values* are just Lisp expressions. So if the value is a string, you need to add the double-quotes around the value yourself.

To control whether an agenda command should be accessible from a specific context, you can customize `org-agenda-custom-commands-contexts`. Let's say for example that you have an agenda command `"o"` displaying a view that you only need when reading emails. Then you would configure this option like this:

```
(setq org-agenda-custom-commands-contexts
      '(("o" (in-mode . "message-mode"))))
```

You can also tell that the command key `"o"` should refer to another command key `"r"`. In that case, add this command key like this:

```
(setq org-agenda-custom-commands-contexts
      '(("o" "r" (in-mode . "message-mode"))))
```

See the docstring of the variable for more information.

10.7 Exporting agenda views

If you are away from your computer, it can be very useful to have a printed version of some agenda views to carry around. Org mode can export custom agenda views as plain text,

HTML[15], Postscript, PDF[16], and iCalendar files. If you want to do this only occasionally, use the command

C-x C-w `org-agenda-write`

Write the agenda view to a file. Depending on the extension of the selected file name, the view will be exported as HTML (extension `.html` or `.htm`), Postscript (extension `.ps`), iCalendar (extension `.ics`), or plain text (any other extension). Use the variable `org-agenda-exporter-settings` to set options for `ps-print` and for `htmlize` to be used during export, for example

```
(setq org-agenda-exporter-settings
      '((ps-number-of-columns 2)
        (ps-landscape-mode t)
        (org-agenda-add-entry-text-maxlines 5)
        (htmlize-output-type 'css)))
```

If you need to export certain agenda views frequently, you can associate any custom agenda command with a list of output file names[17]. Here is an example that first defines custom commands for the agenda and the global TODO list, together with a number of files to which to export them. Then we define two block agenda commands and specify file names for them as well. File names can be relative to the current working directory, or absolute.

```
(setq org-agenda-custom-commands
      '(("X" agenda "" nil ("agenda.html" "agenda.ps"))
        ("Y" alltodo "" nil ("todo.html" "todo.txt" "todo.ps"))
        ("h" "Agenda and Home-related tasks"
         ((agenda "")
          (tags-todo "home")
          (tags "garden"))
         nil
         ("~/views/home.html"))
        ("o" "Agenda and Office-related tasks"
         ((agenda)
          (tags-todo "work")
          (tags "office"))
         nil
         ("~/views/office.ps" "~/calendars/office.ics"))))
```

The extension of the file name determines the type of export. If it is `.html`, Org mode will try to use the `htmlize.el` package to convert the buffer to HTML and save it to this file name. If the extension is `.ps`, `ps-print-buffer-with-faces` is used to produce Postscript output. If the extension is `.ics`, iCalendar export is run export over all files that were used to construct the agenda, and limit the export to entries listed in the agenda. Any other extension produces a plain ASCII file.

---

[15] You need to install `htmlize.el` from Hrvoje Niksic's repository.

[16] To create PDF output, the ghostscript `ps2pdf` utility must be installed on the system. Selecting a PDF file will also create the postscript file.

[17] If you want to store standard views like the weekly agenda or the global TODO list as well, you need to define custom commands for them in order to be able to specify file names.

The export files are *not* created when you use one of those commands interactively because this might use too much overhead. Instead, there is a special command to produce *all* specified files in one step:

C-c a e `org-store-agenda-views`
Export all agenda views that have export file names associated with them.

You can use the options section of the custom agenda commands to also set options for the export commands. For example:

```
(setq org-agenda-custom-commands
      '(("X" agenda ""
         ((ps-number-of-columns 2)
          (ps-landscape-mode t)
          (org-agenda-prefix-format " [ ] ")
          (org-agenda-with-colors nil)
          (org-agenda-remove-tags t))
         ("theagenda.ps"))))
```

This command sets two options for the Postscript exporter, to make it print in two columns in landscape format—the resulting page can be cut in two and then used in a paper agenda. The remaining settings modify the agenda prefix to omit category and scheduling information, and instead include a checkbox to check off items. We also remove the tags to make the lines compact, and we don't want to use colors for the black-and-white printer. Settings specified in `org-agenda-exporter-settings` will also apply, but the settings in `org-agenda-custom-commands` take precedence.

From the command line you may also use

```
emacs -eval (org-batch-store-agenda-views) -kill
```

or, if you need to modify some parameters[18]

```
emacs -eval '(org-batch-store-agenda-views                      \
              org-agenda-span (quote month)                     \
              org-agenda-start-day "2007-11-01"                 \
              org-agenda-include-diary nil                      \
              org-agenda-files (quote ("~/org/project.org")))'  \
        -kill
```

which will create the agenda views restricted to the file `~/org/project.org`, without diary entries and with a 30-day extent.

You can also extract agenda information in a way that allows further processing by other programs. See Section A.10 [Extracting agenda information], page 250, for more information.

10.8 Using column view in the agenda

Column view (see Section 7.5 [Column view], page 67) is normally used to view and edit properties embedded in the hierarchical structure of an Org file. It can be quite useful to use column view also from the agenda, where entries are collected by certain criteria.

[18] Quoting depends on the system you use, please check the FAQ for examples.

C-c C-x C-c **org-agenda-columns**
Turn on column view in the agenda.

To understand how to use this properly, it is important to realize that the entries in the agenda are no longer in their proper outline environment. This causes the following issues:

1. Org needs to make a decision which **COLUMNS** format to use. Since the entries in the agenda are collected from different files, and different files may have different **COLUMNS** formats, this is a non-trivial problem. Org first checks if the variable **org-agenda-overriding-columns-format** is currently set, and if so, takes the format from there. Otherwise it takes the format associated with the first item in the agenda, or, if that item does not have a specific format—defined in a property, or in its file—it uses **org-columns-default-format**.
2. If any of the columns has a summary type defined (see Section 7.5.1.2 [Column attributes], page 68), turning on column view in the agenda will visit all relevant agenda files and make sure that the computations of this property are up to date. This is also true for the special **CLOCKSUM** property. Org will then sum the values displayed in the agenda. In the daily/weekly agenda, the sums will cover a single day; in all other views they cover the entire block. It is vital to realize that the agenda may show the same entry *twice*—for example as scheduled and as a deadline—and it may show two entries from the same hierarchy—for example a *parent* and its *child*. In these cases, the summation in the agenda will lead to incorrect results because some values will count double.
3. When the column view in the agenda shows the **CLOCKSUM**, that is always the entire clocked time for this item. So even in the daily/weekly agenda, the clocksum listed in column view may originate from times outside the current view. This has the advantage that you can compare these values with a column listing the planned total effort for a task—one of the major applications for column view in the agenda. If you want information about clocked time in the displayed period use clock table mode (press *R* in the agenda).
4. When the column view in the agenda shows the **CLOCKSUM_T**, that is always today's clocked time for this item. So even in the weekly agenda, the clocksum listed in column view only originates from today. This lets you compare the time you spent on a task for today, with the time already spent —via **CLOCKSUM**—and with the planned total effort for it.

11 Markup for rich export

When exporting Org mode documents, the exporter tries to reflect the structure of the document as accurately as possible in the back-end. Since export targets like HTML and LaTeX allow much richer formatting, Org mode has rules on how to prepare text for rich export. This section summarizes the markup rules used in an Org mode buffer.

11.1 Paragraphs, line breaks, and quoting

Paragraphs are separated by at least one empty line. If you need to enforce a line break within a paragraph, use '\\' at the end of a line.

To preserve the line breaks, indentation and blank lines in a region, but otherwise use normal formatting, you can use this construct, which can also be used to format poetry.

```
#+BEGIN_VERSE
 Great clouds overhead
 Tiny black birds rise and fall
 Snow covers Emacs

     -- AlexSchroeder
#+END_VERSE
```

When quoting a passage from another document, it is customary to format this as a paragraph that is indented on both the left and the right margin. You can include quotations in Org mode documents like this:

```
#+BEGIN_QUOTE
Everything should be made as simple as possible,
but not any simpler -- Albert Einstein
#+END_QUOTE
```

If you would like to center some text, do it like this:

```
#+BEGIN_CENTER
Everything should be made as simple as possible, \\
but not any simpler
#+END_CENTER
```

11.2 Emphasis and monospace

You can make words `*bold*`, */italic/*, _underlined_, `=verbatim=` and `~code~`, and, if you must, '`+strike-through+`'. Text in the code and verbatim string is not processed for Org mode specific syntax, it is exported verbatim.

To turn off fontification for marked up text, you can set `org-fontify-emphasized-text` to `nil`. To narrow down the list of available markup syntax, you can customize `org-emphasis-alist`. To fine tune what characters are allowed before and after the markup characters, you can tweak `org-emphasis-regexp-components`. Beware that changing one of the above variables will no take effect until you reload Org, for which you may need to restart Emacs.

11.3 Horizontal rules

A line consisting of only dashes, and at least 5 of them, will be exported as a horizontal line.

11.4 Images and Tables

Both the native Org mode tables (see Chapter 3 [Tables], page 19) and tables formatted with the `table.el` package will be exported properly. For Org mode tables, the lines before the first horizontal separator line will become table header lines. You can use the following lines somewhere before the table to assign a caption and a label for cross references, and in the text you can refer to the object with `[[tab:basic-data]]` (see Section 4.2 [Internal links], page 38):

```
#+CAPTION: This is the caption for the next table (or link)
#+NAME:   tab:basic-data
   | ... | ...|
   |-----|----|
```

Optionally, the caption can take the form:

```
#+CAPTION[Caption for list of tables]: Caption for table.
```

Some back-ends allow you to directly include images into the exported document. Org does this, if a link to an image files does not have a description part, for example `[[./img/a.jpg]]`. If you wish to define a caption for the image and maybe a label for internal cross references, make sure that the link is on a line by itself and precede it with `#+CAPTION` and `#+NAME` as follows:

```
#+CAPTION: This is the caption for the next figure link (or table)
#+NAME:   fig:SED-HR4049
[[./img/a.jpg]]
```

Such images can be displayed within the buffer. See Section 4.4 [Handling links], page 41.

Even though images and tables are prominent examples of captioned structures, the same caption mechanism can apply to many others (e.g., LaTeX equations, source code blocks). Depending on the export back-end, those may or may not be handled.

11.5 Literal examples

You can include literal examples that should not be subjected to markup. Such examples will be typeset in monospace, so this is well suited for source code and similar examples.

```
#+BEGIN_EXAMPLE
Some example from a text file.
#+END_EXAMPLE
```

Note that such blocks may be *indented* in order to align nicely with indented text and in particular with plain list structure (see Section 2.7 [Plain lists], page 12). For simplicity when using small examples, you can also start the example lines with a colon followed by a space. There may also be additional whitespace before the colon:

```
Here is an example
   : Some example from a text file.
```

If the example is source code from a programming language, or any other text that can be marked up by font-lock in Emacs, you can ask for the example to look like the fontified Emacs buffer[1]. This is done with the 'src' block, where you also need to specify the name of the major mode that should be used to fontify the example[2], see Section 15.2 [Easy templates], page 228 for shortcuts to easily insert code blocks.

```
#+BEGIN_SRC emacs-lisp
  (defun org-xor (a b)
     "Exclusive or."
     (if a (not b) b))
#+END_SRC
```

Both in `example` and in `src` snippets, you can add a `-n` switch to the end of the `BEGIN` line, to get the lines of the example numbered. The `-n` takes an optional numeric argument specifying the starting line number of the block. If you use a `+n` switch, the numbering from the previous numbered snippet will be continued in the current one. The `+n` can also take a numeric argument. The value of the argument will be added to the last line of the previous block to determine the starting line number.

```
#+BEGIN_SRC emacs-lisp -n 20
 ;; this will export with line number 20
 (message "This is line 21")
#+END_SRC
#+BEGIN_SRC emacs-lisp +n 10
 ;; This will be listed as line 31
 (message "This is line 32")
#+END_SRC
```

In literal examples, Org will interpret strings like '`(ref:name)`' as labels, and use them as targets for special hyperlinks like `[[(name)]]` (i.e., the reference name enclosed in single parenthesis). In HTML, hovering the mouse over such a link will remote-highlight the corresponding code line, which is kind of cool.

You can also add a `-r` switch which *removes* the labels from the source code[3]. With the `-n` switch, links to these references will be labeled by the line numbers from the code listing, otherwise links will use the labels with no parentheses. Here is an example:

```
#+BEGIN_SRC emacs-lisp -n -r
(save-excursion                  (ref:sc)
   (goto-char (point-min)))      (ref:jump)
#+END_SRC
In line [[(sc)]] we remember the current position.  [[(jump)][Line (jump)]]
jumps to point-min.
```

---

[1] This works automatically for the HTML back-end (it requires version 1.34 of the `htmlize.el` package, which you need to install). Fontified code chunks in LaTeX can be achieved using either the listings or the minted package. If you use minted or listing, you must load the packages manually, for example by adding the desired package to `org-latex-packages-alist`. Refer to `org-latex-listings` for details.

[2] Code in '`src`' blocks may also be evaluated either interactively or on export. See Chapter 14 [Working with source code], page 198, for more information on evaluating code blocks.

[3] Adding `-k` to `-n -r` will *keep* the labels in the source code while using line numbers for the links, which might be useful to explain those in an Org mode example code.

Finally, you can use `-i` to preserve the indentation of a specific code block (see Section 14.2 [Editing source code], page 200).

If the syntax for the label format conflicts with the language syntax, use a `-l` switch to change the format, for example '`#+BEGIN_SRC pascal -n -r -l "((%s))"`'. See also the variable `org-coderef-label-format`.

HTML export also allows examples to be published as text areas (see Section 12.9.10 [Text areas in HTML export], page 154).

Because the `#+BEGIN_...` and `#+END_...` patterns need to be added so often, shortcuts are provided using the Easy templates facility (see Section 15.2 [Easy templates], page 228).

C-c ' Edit the source code example at point in its native mode. This works by switching to a temporary buffer with the source code. You need to exit by pressing *C-c '* again[4]. The edited version will then replace the old version in the Org buffer. Fixed-width regions (where each line starts with a colon followed by a space) will be edited using `artist-mode`[5] to allow creating ASCII drawings easily. Using this command in an empty line will create a new fixed-width region.

C-c l Calling `org-store-link` while editing a source code example in a temporary buffer created with *C-c '* will prompt for a label. Make sure that it is unique in the current buffer, and insert it with the proper formatting like '`(ref:label)`' at the end of the current line. Then the label is stored as a link '`(label)`', for retrieval with *C-c C-l*.

11.6 Special symbols

You can use LaTeX-like syntax to insert special symbols—named entities—like '`\alpha`' to indicate the Greek letter, or '`\to`' to indicate an arrow. Completion for these symbols is available, just type '`\`' and maybe a few letters, and press *M-TAB* to see possible completions. If you need such a symbol inside a word, terminate it with a pair of curly brackets. For example

```
Pro tip: Given a circle \Gamma of diameter d, the length of its circumferenc
is \pi{}d.
```

A large number of entities is provided, with names taken from both HTML and LaTeX; you can comfortably browse the complete list from a dedicated buffer using the command `org-entities-help`. It is also possible to provide your own special symbols in the variable `org-entities-user`.

During export, these symbols are transformed into the native format of the exporter back-end. Strings like `\alpha` are exported as `α` in the HTML output, and as `\(\alpha\)` in the LaTeX output. Similarly, `\nbsp` becomes ` ` in HTML and `~` in LaTeX.

Entities may also be used as a may to escape markup in an Org document, e.g., '`\under{}not underlined\under`' exports as '`_not underlined_`'.

---

[4] Upon exit, lines starting with '`*`', '`,*`', '`#+`' and '`,#+`' will get a comma prepended, to keep them from being interpreted by Org as outline nodes or special syntax. These commas will be stripped for editing with *C-c '*, and also for export.

[5] You may select a different-mode with the variable `org-edit-fixed-width-region-mode`.

If you would like to see entities displayed as UTF-8 characters, use the following command[6]:

`C-c C-x \` Toggle display of entities as UTF-8 characters. This does not change the buffer content which remains plain ASCII, but it overlays the UTF-8 character for display purposes only.

In addition to regular entities defined above, Org exports in a special way[7] the following commonly used character combinations: '`\-`' is treated as a shy hyphen, '`--`' and '`---`' are converted into dashes, and '`...`' becomes a compact set of dots.

11.7 Subscripts and superscripts

'`^`' and '`_`' are used to indicate super- and subscripts. To increase the readability of ASCII text, it is not necessary—but OK—to surround multi-character sub- and superscripts with curly braces. Those are, however, mandatory, when more than one word is involved. For example

```
The radius of the sun is R_sun = 6.96 x 10^8 m.  On the other hand, the
radius of Alpha Centauri is R_{Alpha Centauri} = 1.28 x R_{sun}.
```

If you write a text where the underscore is often used in a different context, Org's convention to always interpret these as subscripts can get in your way. Configure the variable `org-use-sub-superscripts` to change this convention. For example, when setting this variable to `{}`, '`a_b`' will not be interpreted as a subscript, but '`a_{b}`' will.

`C-c C-x \` In addition to showing entities as UTF-8 characters, this command will also format sub- and superscripts in a WYSIWYM way.

11.8 Embedded LaTeX

Plain ASCII is normally sufficient for almost all note taking. Exceptions include scientific notes, which often require mathematical symbols and the occasional formula. LaTeX[8] is widely used to typeset scientific documents. Org mode supports embedding LaTeX code into its files, because many academics are used to writing and reading LaTeX source code, and because it can be readily processed to produce pretty output for a number of export back-ends.

11.8.1 LaTeX fragments

Org mode can contain LaTeX math fragments, and it supports ways to process these for several export back-ends. When exporting to LaTeX, the code is left as it is. When exporting to HTML, Org can use either MathJax (see Section 12.9.9 [Math formatting in HTML export], page 154) or transcode the math into images (see see Section 11.8.2 [Previewing LaTeX fragments], page 135).

LaTeX fragments don't need any special marking at all. The following snippets will be identified as LaTeX source code:

---

[6] You can turn this on by default by setting the variable `org-pretty-entities`, or on a per-file base with the `#+STARTUP` option `entitiespretty`.

[7] This behaviour can be disabled with `-` export setting (see Section 12.2 [Export settings], page 138).

[8] LaTeX is a macro system based on Donald E. Knuth's TeX system. Many of the features described here as "LaTeX" are really from TeX, but for simplicity I am blurring this distinction.

- Environments of any kind[9]. The only requirement is that the `\begin` statement appears on a new line, at the beginning of the line or after whitespaces only.
- Text within the usual LaTeX math delimiters. To avoid conflicts with currency specifications, single '`$`' characters are only recognized as math delimiters if the enclosed text contains at most two line breaks, is directly attached to the '`$`' characters with no whitespace in between, and if the closing '`$`' is followed by whitespace or punctuation (parentheses and quotes are considered to be punctuation in this context). For the other delimiters, there is no such restriction, so when in doubt, use '`\(...\)`' as inline math delimiters.

For example:

```
\begin{equation}
x=\sqrt{b}
\end{equation}

If $a^2=b$ and \( b=2 \), then the solution must be
either $$ a=+\sqrt{2} $$ or \[ a=-\sqrt{2} \].
```

LaTeX processing can be configured with the variable `org-export-with-latex`. The default setting is `t` which means MathJax for HTML, and no processing for ASCII and LaTeX back-ends. You can also set this variable on a per-file basis using one of these lines:

`#+OPTIONS: tex:t`	Do the right thing automatically (MathJax)
`#+OPTIONS: tex:nil`	Do not process LaTeX fragments at all
`#+OPTIONS: tex:verbatim`	Verbatim export, for jsMath or so

11.8.2 Previewing LaTeX fragments

If you have a working LaTeX installation and `dvipng`, `dvisvgm` or `convert` installed[10], LaTeX fragments can be processed to produce images of the typeset expressions to be used for inclusion while exporting to HTML (see see Section 11.8.1 [LaTeX fragments], page 134), or for inline previewing within Org mode.

You can customize the variables `org-format-latex-options` and `org-format-latex-header` to influence some aspects of the preview. In particular, the `:scale` (and for HTML export, `:html-scale`) property of the former can be used to adjust the size of the preview images.

C-c C-x C-l
: Produce a preview image of the LaTeX fragment at point and overlay it over the source code. If there is no fragment at point, process all fragments in the current entry (between two headlines). When called with a prefix argument, process the entire subtree. When called with two prefix arguments, or when the cursor is before the first headline, process the entire buffer.

C-c C-c
: Remove the overlay preview images.

---

[9] When MathJax is used, only the environments recognized by MathJax will be processed. When `dvipng` program, `dvisvgm` program or `imagemagick` suite is used to create images, any LaTeX environment will be handled.

[10] These are respectively available at `http://sourceforge.net/projects/dvipng/`, `http://dvisvgm.bplaced.net/` and from the `imagemagick` suite. Choose the converter by setting the variable `org-preview-latex-default-process` accordingly.

You can turn on the previewing of all LaTeX fragments in a file with

```
#+STARTUP: latexpreview
```

To disable it, simply use

```
#+STARTUP: nolatexpreview
```

11.8.3 Using CDLaTeX to enter math

CDLaTeX mode is a minor mode that is normally used in combination with a major LaTeX mode like AUCTeX in order to speed-up insertion of environments and math templates. Inside Org mode, you can make use of some of the features of CDLaTeX mode. You need to install `cdlatex.el` and `texmathp.el` (the latter comes also with AUCTeX) from https://staff.fnwi.uva.nl/c.dominik/Tools/cdlatex. Don't use CDLaTeX mode itself under Org mode, but use the light version `org-cdlatex-mode` that comes as part of Org mode. Turn it on for the current buffer with *M-x org-cdlatex-mode RET*, or for all Org files with

```
(add-hook 'org-mode-hook 'turn-on-org-cdlatex)
```

When this mode is enabled, the following features are present (for more details see the documentation of CDLaTeX mode):

- Environment templates can be inserted with *C-c {*.
- The `TAB` key will do template expansion if the cursor is inside a LaTeX fragment[11]. For example, `TAB` will expand `fr` to `\frac{}{}` and position the cursor correctly inside the first brace. Another `TAB` will get you into the second brace. Even outside fragments, `TAB` will expand environment abbreviations at the beginning of a line. For example, if you write '`equ`' at the beginning of a line and press `TAB`, this abbreviation will be expanded to an `equation` environment. To get a list of all abbreviations, type *M-x cdlatex-command-help RET*.
- Pressing `_` and `^` inside a LaTeX fragment will insert these characters together with a pair of braces. If you use `TAB` to move out of the braces, and if the braces surround only a single character or macro, they are removed again (depending on the variable `cdlatex-simplify-sub-super-scripts`).
- Pressing the grave accent ` followed by a character inserts math macros, also outside LaTeX fragments. If you wait more than 1.5 seconds after the grave accent, a help window will pop up.
- Pressing the apostrophe ' followed by another character modifies the symbol before point with an accent or a font. If you wait more than 1.5 seconds after the apostrophe, a help window will pop up. Character modification will work only inside LaTeX fragments; outside the quote is normal.

[11] Org mode has a method to test if the cursor is inside such a fragment, see the documentation of the function `org-inside-LaTeX-fragment-p`.

12 Exporting

Sometimes, you may want to pretty print your notes, publish them on the web or even share them with people not using Org. In these cases, the Org export facilities can be used to convert your documents to a variety of other formats, while retaining as much structure (see Chapter 2 [Document structure], page 6) and markup (see Chapter 11 [Markup], page 130) as possible.

Libraries responsible for such translation are called back-ends. Org ships with the following ones

- ascii (ASCII format)
- beamer (LaTeX Beamer format)
- html (HTML format)
- icalendar (iCalendar format)
- latex (LaTeX format)
- md (Markdown format)
- odt (OpenDocument Text format)
- org (Org format)
- texinfo (Texinfo format)
- man (Man page format)

Org also uses additional libraries located in `contrib/` directory (see Section 1.2 [Installation], page 2). Users can install additional export libraries for additional formats from the Emacs packaging system. For easy discovery, these packages have a common naming scheme: `ox-NAME`, where NAME is one of the formats. For example, `ox-koma-letter` for `koma-letter` back-end.

Org loads back-ends for the following formats by default: `ascii`, `html`, `icalendar`, `latex` and `odt`.

Org can load additional back-ends either of two ways: through the `org-export-backends` variable configuration; or, by requiring the library in the Emacs init file like this:

```
(require 'ox-md)
```

12.1 The export dispatcher

The export dispatcher is the main interface for Org's exports. A hierarchical menu presents the currently configured export formats. Options are shown as easy toggle switches on the same screen.

Org also has a minimal prompt interface for the export dispatcher. When the variable `org-export-dispatch-use-expert-ui` is set to a non-`nil` value, Org prompts in the minibuffer. To switch back to the hierarchical menu, press `?`.

C-c C-e `org-export-dispatch`

Invokes the export dispatcher interface. The options show default settings. The *C-u* prefix argument preserves options from the previous export, including any sub-tree selections.

Org exports the entire buffer by default. If the Org buffer has an active region, then Org exports just that region.

These are the export options, the key combinations that toggle them (see Section 12.2 [Export settings], page 138):

`C-a`
: Toggles asynchronous export. Asynchronous export uses an external Emacs process with a specially configured initialization file to complete the exporting process in the background thereby releasing the current interface. This is particularly useful when exporting long documents.

  Output from an asynchronous export is saved on the "the export stack". To view this stack, call the export dispatcher with a double `C-u` prefix argument. If already in the export dispatcher menu, `&` displays the stack.

  To make the background export process the default, customize the variable, `org-export-in-background`.

`C-b`
: Toggle body-only export. Useful for excluding headers and footers in the export. Affects only those back-end formats that have such sections—like `<head>...</head>` in HTML.

`C-s`
: Toggle sub-tree export. When turned on, Org exports only the sub-tree starting from the cursor position at the time the export dispatcher was invoked. Org uses the top heading of this sub-tree as the document's title. If the cursor is not on a heading, Org uses the nearest enclosing header. If the cursor is in the document preamble, Org signals an error and aborts export.

  To make the sub-tree export the default, customize the variable, `org-export-initial-scope`.

`C-v`
: Toggle visible-only export. Useful for exporting only visible parts of an Org document by adjusting outline visibility settings.

12.2 Export settings

Export options can be set: globally with variables; for an individual file by making variables buffer-local with in-buffer settings (see Section 15.6 [In-buffer settings], page 230), by setting individual keywords, or by specifying them in a compact form with the `#+OPTIONS` keyword; or for a tree by setting properties (see Chapter 7 [Properties and columns], page 64). Options set at a specific level override options set at a more general level.

In-buffer settings may appear anywhere in the file, either directly or indirectly through a file included using '`#+SETUPFILE: filename or URL`' syntax. Option keyword sets tailored to a particular back-end can be inserted from the export dispatcher (see Section 12.1 [The export dispatcher], page 137) using the `Insert template` command by pressing `#`. To insert keywords individually, a good way to make sure the keyword is correct is to type `#+` and then to use `M-TAB`[1] for completion.

The export keywords available for every back-end, and their equivalent global variables, include:

'`AUTHOR`'
: The document author (`user-full-name`).

---

[1] Many desktops intercept `M-TAB` to switch windows. Use `C-M-i` or `ESC TAB` instead.

'CREATOR' Entity responsible for output generation (org-export-creator-string).

'DATE' A date or a time-stamp[2].

'EMAIL' The email address (user-mail-address).

'LANGUAGE'
: Language to use for translating certain strings (org-export-default-language). With '#+LANGUAGE: fr', for example, Org translates *Table of contents* to the French *Table des matières*.

'SELECT_TAGS'
: The default value is :export:. When a tree is tagged with :export: (org-export-select-tags), Org selects that tree and its sub-trees for export. Org excludes trees with :noexport: tags, see below. When selectively exporting files with :export: tags set, Org does not export any text that appears before the first headline.

'EXCLUDE_TAGS'
: The default value is :noexport:. When a tree is tagged with :noexport: (org-export-exclude-tags), Org excludes that tree and its sub-trees from export. Entries tagged with :noexport: will be unconditionally excluded from the export, even if they have an :export: tag. Even if a sub-tree is not exported, Org will execute any code blocks contained in them.

'TITLE' Org displays this title. For long titles, use multiple #+TITLE lines.

'EXPORT_FILE_NAME'
: The name of the output file to be generated. Otherwise, Org generates the file name based on the buffer name and the extension based on the back-end format.

The #+OPTIONS keyword is a compact form. To configure multiple options, use several #+OPTIONS lines. #+OPTIONS recognizes the following arguments.

': Toggle smart quotes (org-export-with-smart-quotes). Depending on the language used, when activated, Org treats pairs of double quotes as primary quotes, pairs of single quotes as secondary quotes, and single quote marks as apostrophes.

*: Toggle emphasized text (org-export-with-emphasize).

-: Toggle conversion of special strings (org-export-with-special-strings).

:: Toggle fixed-width sections (org-export-with-fixed-width).

<: Toggle inclusion of time/date active/inactive stamps (org-export-with-timestamps).

\n: Toggles whether to preserve line breaks (org-export-preserve-breaks).

^: Toggle TEX-like syntax for sub- and superscripts. If you write "^:{}", 'a_{b}' will be interpreted, but the simple 'a_b' will be left as it is (org-export-with-sub-superscripts).

[2] The variable org-export-date-timestamp-format defines how this time-stamp will be exported.

`arch:` Configure how archived trees are exported. When set to `headline`, the export process skips the contents and processes only the headlines (`org-export-with-archived-trees`).

`author:` Toggle inclusion of author name into exported file (`org-export-with-author`).

`broken-links:` Toggles if Org should continue exporting upon finding a broken internal link. When set to `mark`, Org clearly marks the problem link in the output (`org-export-with-broken-links`).

`c:` Toggle inclusion of CLOCK keywords (`org-export-with-clocks`).

`creator:` Toggle inclusion of creator information in the exported file (`org-export-with-creator`).

`d:` Toggles inclusion of drawers, or list of drawers to include, or list of drawers to exclude (`org-export-with-drawers`).

`date:` Toggle inclusion of a date into exported file (`org-export-with-date`).

`e:` Toggle inclusion of entities (`org-export-with-entities`).

`email:` Toggle inclusion of the author's e-mail into exported file (`org-export-with-email`).

`f:` Toggle the inclusion of footnotes (`org-export-with-footnotes`).

`H:` Set the number of headline levels for export (`org-export-headline-levels`). Below that level, headlines are treated differently. In most back-ends, they become list items.

`inline:` Toggle inclusion of inlinetasks (`org-export-with-inlinetasks`).

`num:` Toggle section-numbers (`org-export-with-section-numbers`). When set to number '`n`', Org numbers only those headlines at level '`n`' or above. Setting `UNNUMBERED` property to non-`nil` disables numbering of a heading. Since sub-headings inherit from this property, it affects their numbering, too.

`p:` Toggle export of planning information (`org-export-with-planning`). "Planning information" comes from lines located right after the headline and contain any combination of these cookies: `SCHEDULED:`, `DEADLINE:`, or `CLOSED:`.

`pri:` Toggle inclusion of priority cookies (`org-export-with-priority`).

`prop:` Toggle inclusion of property drawers, or list the properties to include (`org-export-with-properties`).

`stat:` Toggle inclusion of statistics cookies (`org-export-with-statistics-cookies`).

`tags:` Toggle inclusion of tags, may also be `not-in-toc` (`org-export-with-tags`).

`tasks:` Toggle inclusion of tasks (TODO items); or `nil` to remove all tasks; or `todo` to remove DONE tasks; or list the keywords to keep (`org-export-with-tasks`).

`tex:` `nil` does not export; `t` exports; `verbatim` keeps everything in verbatim (`org-export-with-latex`).

`timestamp:`
: Toggle inclusion of the creation time in the exported file (`org-export-time-stamp-file`).

`title:`
: Toggle inclusion of title (`org-export-with-title`).

`toc:`
: Toggle inclusion of the table of contents, or set the level limit (`org-export-with-toc`).

`todo:`
: Toggle inclusion of TODO keywords into exported text (`org-export-with-todo-keywords`).

`|:`
: Toggle inclusion of tables (`org-export-with-tables`).

When exporting sub-trees, special node properties in them can override the above keywords. They are special because they have an '`EXPORT_`' prefix. For example, '`DATE`' and '`EXPORT_FILE_NAME`' keywords become, respectively, '`EXPORT_DATE`' and '`EXPORT_FILE_NAME`'. Except for '`SETUPFILE`', all other keywords listed above have an '`EXPORT_`' equivalent.

If `org-export-allow-bind-keywords` is non-`nil`, Emacs variables can become buffer-local during export by using the BIND keyword. Its syntax is '`#+BIND: variable value`'. This is particularly useful for in-buffer settings that cannot be changed using keywords.

12.3 Table of contents

Org normally inserts the table of contents directly before the first headline of the file. Org sets the TOC depth the same as the headline levels in the file. Use a lower number for lower TOC depth. To turn off TOC entirely, use `nil`. This is configured in the `org-export-with-toc` variable or as keywords in an Org file as:

```
#+OPTIONS: toc:2             only include two levels in TOC
#+OPTIONS: toc:nil           no default TOC at all
```

To move the table of contents to a different location, first turn off the default with `org-export-with-toc` variable or with `#+OPTIONS: toc:nil`. Then insert `#+TOC: headlines N` at the desired location(s).

```
#+OPTIONS: toc:nil           no default TOC
...
#+TOC: headlines 2           insert TOC here, with two headline levels
```

To adjust the TOC depth for a specific section of the Org document, append an additional '`local`' parameter. This parameter becomes a relative depth for the current level.

Note that for this feature to work properly in LaTeX export, the Org file requires the inclusion of the `titletoc` package. Because of compatibility issues, `titletoc` has to be loaded *before* `hyperref`. Customize the `org-latex-default-packages-alist` variable.

```
* Section #+TOC: headlines 1 local insert local TOC, with direct children
only
```

Use the `TOC` keyword to generate list of tables (resp. all listings) with captions.

```
#+TOC: listings              build a list of listings
#+TOC: tables                build a list of tables
```

Normally Org uses the headline for its entry in the table of contents. But with `ALT_TITLE` property, a different entry can be specified for the table of contents.

12.4 Include files

Include other files during export. For example, to include your `.emacs` file, you could use:

```
#+INCLUDE: "~/.emacs" src emacs-lisp
```

The first parameter is the file name to include. The optional second parameter specifies the block type: '`example`', '`export`' or '`src`'. The optional third parameter specifies the source code language to use for formatting the contents. This is relevant to both '`export`' and '`src`' block types.

If an include file is specified as having a markup language, Org neither checks for valid syntax nor changes the contents in any way. For '`example`' and '`src`' blocks, Org code-escapes the contents before inclusion.

If an include file is not specified as having any markup language, Org assumes it be in Org format and proceeds as usual with a few exceptions. Org makes the footnote labels (see Section 2.10 [Footnotes], page 16) in the included file local to that file. The contents of the included file will belong to the same structure—headline, item—containing the `INCLUDE` keyword. In particular, headlines within the file will become children of the current section. That behavior can be changed by providing an additional keyword parameter, `:minlevel`. It shifts the headlines in the included file to become the lowest level. For example, this syntax makes the included file a sibling of the current top-level headline:

```
#+INCLUDE: "~/my-book/chapter2.org" :minlevel 1
```

Inclusion of only portions of files are specified using ranges parameter with `:lines` keyword. The line at the upper end of the range will not be included. The start and/or the end of the range may be omitted to use the obvious defaults.

`#+INCLUDE: "~/.emacs" :lines "5-10"`	Include lines 5 to 10, 10 excluded
`#+INCLUDE: "~/.emacs" :lines "-10"`	Include lines 1 to 10, 10 excluded
`#+INCLUDE: "~/.emacs" :lines "10-"`	Include lines from 10 to EOF

Inclusions may specify a file-link to extract an object matched by `org-link-search`[3] (see Section 4.7 [Search options], page 45).

To extract only the contents of the matched object, set `:only-contents` property to non-`nil`. This will omit any planning lines or property drawers. The ranges for `:lines` keyword are relative to the requested element. Some examples:

`#+INCLUDE: "./paper.org::#theory" :only-contents t`
    Include the body of the heading with the custom id '`theory`'
`#+INCLUDE: "./paper.org::mytable"` Include named element.
`#+INCLUDE: "./paper.org::*conclusion" :lines 1-20`
    Include the first 20 lines of the headline named '`conclusion`'.

C-c '   Visit the include file at point.

12.5 Macro replacement

Macros replace text snippets during export. Macros are defined globally in `org-export-global-macros`, or document-wise with the following syntax:

[3] Note that `org-link-search-must-match-exact-headline` is locally bound to non-`nil`. Therefore, `org-link-search` only matches headlines and named elements.

```
#+MACRO: name   replacement text $1, $2 are arguments
```

which can be referenced using `{{{name(arg1, arg2)}}}`[4].

Org recognizes macro references in following Org markup areas: paragraphs, headlines, verse blocks, tables cells and lists. Org also recognizes macro references in keywords, such as `#+CAPTION`, `#+TITLE`, `#+AUTHOR`, `#+DATE`, and for some back-end specific export options.

Org comes with following pre-defined macros:

`{{{title}}}`
`{{{author}}}`
`{{{email}}}`
: Org replaces these macro references with available information at the time of export.

`{{{date}}}`
`{{{date(FORMAT)}}}`
: This macro refers to the `#+DATE` keyword. *FORMAT* is an optional argument to the `{{{date}}}` macro that will be used only if `#+DATE` is a single timestamp. *FORMAT* should be a format string understood by `format-time-string`.

`{{{time(FORMAT)}}}`
`{{{modification-time(FORMAT, VC)}}}`
: These macros refer to the document's date and time of export and date and time of modification. *FORMAT* is a string understood by `format-time-string`. If the second argument to the `modification-time` macro is non-`nil`, Org uses `vc.el` to retrieve the document's modification time from the version control system. Otherwise Org reads the file attributes.

`{{{input-file}}}`
: This macro refers to the filename of the exported file.

`{{{property(PROPERTY-NAME)}}}`
`{{{property(PROPERTY-NAME,SEARCH-OPTION)}}}`
: This macro returns the value of property *PROPERTY-NAME* in the current entry. If *SEARCH-OPTION* (see Section 4.7 [Search options], page 45) refers to a remote entry, that will be used instead.

`{{{n}}}`
`{{{n(NAME)}}}`
`{{{n(NAME,ACTION)}}}`
: This macro implements custom counters by returning the number of times the macro has been expanded so far while exporting the buffer. You can create more than one counter using different *NAME* values. If *ACTION* is `-`, previous value of the counter is held, i.e. the specified counter is not incremented. If the value is a number, the specified counter is set to that value. If it is any other non-empty string, the specified counter is reset to 1. You may leave *NAME* empty to reset the default counter.

---

[4] Since commas separate the arguments, commas within arguments have to be escaped with the backslash character. So only those backslash characters before a comma need escaping with another backslash character.

The surrounding brackets can be made invisible by setting `org-hide-macro-markers` non-`nil`.

Org expands macros at the very beginning of the export process.

12.6 Comment lines

Lines starting with zero or more whitespace characters followed by one '`#`' and a whitespace are treated as comments and, as such, are not exported.

Likewise, regions surrounded by '`#+BEGIN_COMMENT`' ... '`#+END_COMMENT`' are not exported.

Finally, a '`COMMENT`' keyword at the beginning of an entry, but after any other keyword or priority cookie, comments out the entire subtree. In this case, the subtree is not exported and no code block within it is executed either[5]. The command below helps changing the comment status of a headline.

C-c ; Toggle the '`COMMENT`' keyword at the beginning of an entry.

12.7 ASCII/Latin-1/UTF-8 export

ASCII export produces an output file containing only plain ASCII characters. This is the most simplest and direct text output. It does not contain any Org markup either. Latin-1 and UTF-8 export use additional characters and symbols available in these encoding standards. All three of these export formats offer the most basic of text output for maximum portability.

On export, Org fills and justifies text according to the text width set in `org-ascii-text-width`.

Org exports links using a footnote-like style where the descriptive part is in the text and the link is in a note before the next heading. See the variable `org-ascii-links-to-notes` for details.

ASCII export commands

C-c C-e t a/l/u `org-ascii-export-to-ascii`
: Export as an ASCII file with a `.txt` extension. For `myfile.org`, Org exports to `myfile.txt`, overwriting without warning. For `myfile.txt`, Org exports to `myfile.txt.txt` in order to prevent data loss.

C-c C-e t A/L/U `org-ascii-export-as-ascii`
: Export to a temporary buffer. Does not create a file.

ASCII specific export settings

The ASCII export back-end has one extra keyword for customizing ASCII output. Setting this keyword works similar to the general options (see Section 12.2 [Export settings], page 138).

[5] For a less drastic behavior, consider using a select tag (see Section 12.2 [Export settings], page 138) instead.

'SUBTITLE'
: The document subtitle. For long subtitles, use multiple #+SUBTITLE lines in the Org file. Org prints them on one continuous line, wrapping into multiple lines if necessary.

Header and sectioning structure

Org converts the first three outline levels into headlines for ASCII export. The remaining levels are turned into lists. To change this cut-off point where levels become lists, see Section 12.2 [Export settings], page 138.

Quoting ASCII text

To insert text within the Org file by the ASCII back-end, use one the following constructs, inline, keyword, or export block:

```
Inline text @@ascii:and additional text@@ within a paragraph.

#+ASCII: Some text

#+BEGIN_EXPORT ascii
Org exports text in this block only when using ASCII back-end.
#+END_EXPORT
```

ASCII specific attributes

ASCII back-end recognizes only one attribute, :width, which specifies the width of an horizontal rule in number of characters. The keyword and syntax for specifying widths is:

```
#+ATTR_ASCII: :width 10
-----
```

ASCII special blocks

Besides #+BEGIN_CENTER blocks (see Section 11.1 [Paragraphs], page 130), ASCII back-end has these two left and right justification blocks:

```
#+BEGIN_JUSTIFYLEFT
It's just a jump to the left...
#+END_JUSTIFYLEFT

#+BEGIN_JUSTIFYRIGHT
...and then a step to the right.
#+END_JUSTIFYRIGHT
```

12.8 Beamer export

Org uses *Beamer* export to convert an Org file tree structure into a high-quality interactive slides for presentations. *Beamer* is a LaTeX document class for creating presentations in PDF, HTML, and other popular display formats.

12.8.1 Beamer export commands

C-c C-e l b `org-beamer-export-to-latex`
: Export as LaTeX file with a `.tex` extension. For `myfile.org`, Org exports to `myfile.tex`, overwriting without warning.

C-c C-e l B `org-beamer-export-as-latex`
: Export to a temporary buffer. Does not create a file.

C-c C-e l P `org-beamer-export-to-pdf`
: Export as LaTeX file and then convert it to PDF format.

C-c C-e l O
: Export as LaTeX file, convert it to PDF format, and then open the PDF file.

12.8.2 Beamer specific export settings

Beamer export back-end has several additional keywords for customizing Beamer output. These keywords work similar to the general options settings (see Section 12.2 [Export settings], page 138).

'BEAMER_THEME'
: The Beamer layout theme (`org-beamer-theme`). Use square brackets for options. For example:

```
#+BEAMER_THEME: Rochester [height=20pt]
```

'BEAMER_FONT_THEME'
: The Beamer font theme.

'BEAMER_INNER_THEME'
: The Beamer inner theme.

'BEAMER_OUTER_THEME'
: The Beamer outer theme.

'BEAMER_HEADER'
: Arbitrary lines inserted in the preamble, just before the 'hyperref' settings.

'DESCRIPTION'
: The document description. For long descriptions, use multiple `#+DESCRIPTION` keywords. By default, 'hyperref' inserts `#+DESCRIPTION` as metadata. Use `org-latex-hyperref-template` to configure document metadata. Use `org-latex-title-command` to configure typesetting of description as part of front matter.

'KEYWORDS'
: The keywords for defining the contents of the document. Use multiple `#+KEYWORDS` lines if necessary. By default, 'hyperref' inserts `#+KEYWORDS` as metadata. Use `org-latex-hyperref-template` to configure document metadata. Use `org-latex-title-command` to configure typesetting of keywords as part of front matter.

'SUBTITLE'
: Document's subtitle. For typesetting, use `org-beamer-subtitle-format` string. Use `org-latex-hyperref-template` to configure document metadata.

Use `org-latex-title-command` to configure typesetting of subtitle as part of front matter.

12.8.3 Sectioning, Frames and Blocks in Beamer

Org transforms heading levels into Beamer's sectioning elements, frames and blocks. Any Org tree with a not-too-deep-level nesting should in principle be exportable as a Beamer presentation.

- Org headlines become Beamer frames when the heading level in Org is equal to `org-beamer-frame-level` or `H` value in an `OPTIONS` line (see Section 12.2 [Export settings], page 138).

  Org overrides headlines to frames conversion for the current tree of an Org file if it encounters the `BEAMER_ENV` property set to `frame` or `fullframe`. Org ignores whatever `org-beamer-frame-level` happens to be for that headline level in the Org tree. In Beamer terminology, a `fullframe` is a frame without its title.
- Org exports a Beamer frame's objects as `block` environments. Org can enforce wrapping in special block types when `BEAMER_ENV` property is set[6]. For valid values see `org-beamer-environments-default`. To add more values, see `org-beamer-environments-extra`.
- If `BEAMER_ENV` is set to `appendix`, Org exports the entry as an appendix. When set to `note`, Org exports the entry as a note within the frame or between frames, depending on the entry's heading level. When set to `noteNH`, Org exports the entry as a note without its title. When set to `againframe`, Org exports the entry with `\againframe` command, which makes setting the `BEAMER_REF` property mandatory because `\againframe` needs frame to resume.

  When `ignoreheading` is set, Org export ignores the entry's headline but not its content. This is useful for inserting content between frames. It is also useful for properly closing a `column` environment.

When `BEAMER_ACT` is set for a headline, Org export translates that headline as an overlay or action specification. When enclosed in square brackets, Org export makes the overlay specification a default. Use `BEAMER_OPT` to set any options applicable to the current Beamer frame or block. The Beamer export back-end wraps with appropriate angular or square brackets. It also adds the `fragile` option for any code that may require a verbatim block.

To create a column on the Beamer slide, use the `BEAMER_COL` property for its headline in the Org file. Set the value of `BEAMER_COL` to a decimal number representing the fraction of the total text width. Beamer export uses this value to set the column's width and fills the column with the contents of the Org entry. If the Org entry has no specific environment defined, Beamer export ignores the heading. If the Org entry has a defined environment, Beamer export uses the heading as title. Behind the scenes, Beamer export automatically handles LaTeX column separations for contiguous headlines. To manually adjust them for any unique configurations needs, use the `BEAMER_ENV` property.

---

[6] If `BEAMER_ENV` is set, Org export adds `:B_environment:` tag to make it visible. The tag serves as a visual aid and has no semantic relevance.

12.8.4 Beamer specific syntax

Since Org's Beamer export back-end is an extension of the LaTeX back-end, it recognizes other LaTeX specific syntax—for example, '#+LATEX:' or '#+ATTR_LATEX:'. See Section 12.10 [LaTeX export], page 157, for details.

Beamer export wraps the table of contents generated with toc:t OPTION keyword in a frame environment. Beamer export does not wrap the table of contents generated with TOC keyword (see Section 12.3 [Table of contents], page 141). Use square brackets for specifying options.

```
#+TOC: headlines [currentsection]
```

Insert Beamer-specific code using the following constructs:

```
#+BEAMER: \pause

#+BEGIN_EXPORT beamer
Only Beamer export back-end will export this line.
#+END_BEAMER

Text @@beamer:some code@@ within a paragraph.
```

Inline constructs, such as the last one above, are useful for adding overlay specifications to objects with bold, item, link, radio-target and target types. Enclose the value in angular brackets and place the specification at the beginning the object as shown in this example:

```
A *@@beamer:<2->@@useful* feature
```

Beamer export recognizes the ATTR_BEAMER keyword with the following attributes from Beamer configurations: :environment for changing local Beamer environment, :overlay for specifying Beamer overlays in angular or square brackets, and :options for inserting optional arguments.

```
#+ATTR_BEAMER: :environment nonindentlist
- item 1, not indented
- item 2, not indented
- item 3, not indented

#+ATTR_BEAMER: :overlay <+->
- item 1
- item 2

#+ATTR_BEAMER: :options [Lagrange]
Let $G$ be a finite group, and let $H$ be
a subgroup of $G$.  Then the order of $H$ divides the order of $G$.
```

12.8.5 Editing support

The org-beamer-mode is a special minor mode for faster editing of Beamer documents.

```
#+STARTUP: beamer
```

C-c C-b org-beamer-select-environment
: The org-beamer-mode provides this key for quicker selections in Beamer normal environments, and for selecting the BEAMER_COL property.

12.8.6 A Beamer example

Here is an example of an Org document ready for Beamer export.

```
#+TITLE: Example Presentation
#+AUTHOR: Carsten Dominik
#+OPTIONS: H:2 toc:t num:t
#+LATEX_CLASS: beamer
#+LATEX_CLASS_OPTIONS: [presentation]
#+BEAMER_THEME: Madrid
#+COLUMNS: %45ITEM %10BEAMER_ENV(Env) %10BEAMER_ACT(Act) %4BEAMER_COL(Col) %

* This is the first structural section

** Frame 1
*** Thanks to Eric Fraga                                           :B_block:
    :PROPERTIES:
    :BEAMER_COL: 0.48
    :BEAMER_ENV: block
    :END:
    for the first viable Beamer setup in Org
*** Thanks to everyone else                                        :B_block:
    :PROPERTIES:
    :BEAMER_COL: 0.48
    :BEAMER_ACT: <2->
    :BEAMER_ENV: block
    :END:
    for contributing to the discussion
**** This will be formatted as a beamer note                       :B_note:
     :PROPERTIES:
     :BEAMER_env: note
     :END:
** Frame 2 (where we will not use columns)
*** Request
    Please test this stuff!
```

12.9 HTML export

Org mode contains an HTML exporter with extensive HTML formatting compatible with XHTML 1.0 strict standard.

12.9.1 HTML export commands

C-c C-e h h `org-html-export-to-html`
: Export as HTML file with a `.html` extension. For `myfile.org`, Org exports to `myfile.html`, overwriting without warning. *C-c C-e h o* Exports to HTML and opens it in a web browser.

C-c C-e h H `org-html-export-as-html`
: Exports to a temporary buffer. Does not create a file.

12.9.2 HTML Specific export settings

HTML export has a number of keywords, similar to the general options settings described in Section 12.2 [Export settings], page 138.

'DESCRIPTION'
: This is the document's description, which the HTML exporter inserts it as a HTML meta tag in the HTML file. For long descriptions, use multiple `#+DESCRIPTION` lines. The exporter takes care of wrapping the lines properly.

'HTML_DOCTYPE'
: Specify the document type, for example: HTML5 (`org-html-doctype`).

'HTML_CONTAINER'
: Specify the HTML container, such as '`div`', for wrapping sections and elements (`org-html-container-element`).

'HTML_LINK_HOME'
: The URL for home link (`org-html-link-home`).

'HTML_LINK_UP'
: The URL for the up link of exported HTML pages (`org-html-link-up`).

'HTML_MATHJAX'
: Options for MathJax (`org-html-mathjax-options`). MathJax is used to typeset LaTeX math in HTML documents. See Section 12.9.9 [Math formatting in HTML export], page 154, for an example.

'HTML_HEAD'
: Arbitrary lines for appending to the HTML document's head (`org-html-head`).

'HTML_HEAD_EXTRA'
: More arbitrary lines for appending to the HTML document's head (`org-html-head-extra`).

'KEYWORDS'
: Keywords to describe the document's content. HTML exporter inserts these keywords as HTML meta tags. For long keywords, use multiple `#+KEYWORDS` lines.

'LATEX_HEADER'
: Arbitrary lines for appending to the preamble; HTML exporter appends when transcoding LaTeX fragments to images (see Section 12.9.9 [Math formatting in HTML export], page 154).

'SUBTITLE'
: The document's subtitle. HTML exporter formats subtitle if document type is '`HTML5`' and the CSS has a '`subtitle`' class.

Some of these keywords are explained in more detail in the following sections of the manual.

12.9.3 HTML doctypes

Org can export to various (X)HTML flavors.

Set the `org-html-doctype` variable for different (X)HTML variants. Depending on the variant, the HTML exporter adjusts the syntax of HTML conversion accordingly. Org includes the following ready-made variants:

- "html4-strict"
- "html4-transitional"
- "html4-frameset"
- "xhtml-strict"
- "xhtml-transitional"
- "xhtml-frameset"
- "xhtml-11"
- "html5"
- "xhtml5"

See the variable `org-html-doctype-alist` for details. The default is "xhtml-strict".

Org's HTML exporter does not by default enable new block elements introduced with the HTML5 standard. To enable them, set `org-html-html5-fancy` to non-nil. Or use an `OPTIONS` line in the file to set `html5-fancy`. HTML5 documents can now have arbitrary `#+BEGIN` and `#+END` blocks. For example:

```
#+BEGIN_aside
Lorem ipsum
#+END_aside
```

Will export to:

```
<aside>
  <p>Lorem ipsum</p>
</aside>
```

While this:

```
#+ATTR_HTML: :controls controls :width 350
#+BEGIN_video
#+HTML: <source src="movie.mp4" type="video/mp4">
#+HTML: <source src="movie.ogg" type="video/ogg">
Your browser does not support the video tag.
#+END_video
```

Exports to:

```
<video controls="controls" width="350">
  <source src="movie.mp4" type="video/mp4">
  <source src="movie.ogg" type="video/ogg">
  <p>Your browser does not support the video tag.</p>
</video>
```

When special blocks do not have a corresponding HTML5 element, the HTML exporter reverts to standard translation (see `org-html-html5-elements`). For example, `#+BEGIN_lederhosen` exports to '`<div class="lederhosen">`'.

Special blocks cannot have headlines. For the HTML exporter to wrap the headline and its contents in '<section>' or '<article>' tags, set the HTML_CONTAINER property for the headline.

12.9.4 HTML preamble and postamble

The HTML exporter has delineations for preamble and postamble. The default value for org-html-preamble is t, which makes the HTML exporter insert the preamble. See the variable org-html-preamble-format for the format string.

Set org-html-preamble to a string to override the default format string. If the string is a function, the HTML exporter expects the function to return a string upon execution. The HTML exporter inserts this string in the preamble. The HTML exporter will not insert a preamble if org-html-preamble is set nil.

The default value for org-html-postamble is auto, which makes the HTML exporter build a postamble from looking up author's name, email address, creator's name, and date. Set org-html-postamble to t to insert the postamble in the format specified in the org-html-postamble-format variable. The HTML exporter will not insert a postamble if org-html-postamble is set to nil.

12.9.5 Quoting HTML tags

The HTML export back-end transforms '<' and '>' to '<' and '>'. To include raw HTML code in the Org file so the HTML export back-end can insert that HTML code in the output, use this inline syntax: '@@html:'. For example: '@@html:<b>@@bold text@@html:</b>@@'. For larger raw HTML code blocks, use these HTML export code blocks:

```
#+HTML: Literal HTML code for export
```

or

```
#+BEGIN_EXPORT html
All lines between these markers are exported literally
#+END_EXPORT
```

12.9.6 Links in HTML export

The HTML export back-end transforms Org's internal links (see Section 4.2 [Internal links], page 38) to equivalent HTML links in the output. The back-end similarly handles Org's automatic links created by radio targets (see Section 4.2.1 [Radio targets], page 39) similarly. For Org links to external files, the back-end transforms the links to *relative* paths.

For Org links to other .org files, the back-end automatically changes the file extension to .html and makes file paths relative. If the .org files have an equivalent .html version at the same location, then the converted links should work without any further manual intervention. However, to disable this automatic path translation, set org-html-link-org-files-as-html to nil. When disabled, the HTML export back-end substitutes the 'id:'-based links in the HTML output. For more about linking files when publishing to a directory, see Section 13.1.6 [Publishing links], page 193.

Org files can also have special directives to the HTML export back-end. For example, by using #+ATTR_HTML lines to specify new format attributes to <a> or <img> tags. This example shows changing the link's title and style:

```
#+ATTR_HTML: :title The Org mode homepage :style color:red;
[[http://orgmode.org]]
```

12.9.7 Tables in HTML export

The HTML export back-end uses `org-html-table-default-attributes` when exporting Org tables to HTML. By default, the exporter does not draw frames and cell borders. To change for this for a table, use the following lines before the table in the Org file:

```
#+CAPTION: This is a table with lines around and between cells
#+ATTR_HTML: :border 2 :rules all :frame border
```

The HTML export back-end preserves column groupings in Org tables (see Section 3.3 [Column groups], page 23) when exporting to HTML.

Additional options for customizing tables for HTML export.

`org-html-table-align-individual-fields`
: Non-`nil` attaches style attributes for alignment to each table field.

`org-html-table-caption-above`
: Non-`nil` places caption string at the beginning of the table.

`org-html-table-data-tags`
: Opening and ending tags for table data fields.

`org-html-table-default-attributes`
: Default attributes and values for table tags.

`org-html-table-header-tags`
: Opening and ending tags for table's header fields.

`org-html-table-row-tags`
: Opening and ending tags for table rows.

`org-html-table-use-header-tags-for-first-column`
: Non-`nil` formats column one in tables with header tags.

12.9.8 Images in HTML export

The HTML export back-end has features to convert Org image links to HTML inline images and HTML clickable image links.

When the link in the Org file has no description, the HTML export back-end by default in-lines that image. For example: '`[[file:myimg.jpg]]`' is in-lined, while '`[[file:myimg.jpg][the image]]`' links to the text, '`the image`'.

For more details, see the variable `org-html-inline-images`.

On the other hand, if the description part of the Org link is itself another link, such as `file:` or `http:` URL pointing to an image, the HTML export back-end in-lines this image and links to the main image. This Org syntax enables the back-end to link low-resolution thumbnail to the high-resolution version of the image, as shown in this example:

```
[[file:highres.jpg][file:thumb.jpg]]
```

To change attributes of in-lined images, use `#+ATTR_HTML` lines in the Org file. This example shows realignment to right, and adds `alt` and `title` attributes in support of text viewers and modern web accessibility standards.

```
#+CAPTION: A black cat stalking a spider
#+ATTR_HTML: :alt cat/spider image :title Action! :align right
[[./img/a.jpg]]
```

The HTML export back-end copies the `http` links from the Org file as is.

12.9.9 Math formatting in HTML export

LaTeX math snippets (see Section 11.8.1 [LaTeX fragments], page 134) can be displayed in two different ways on HTML pages. The default is to use MathJax which should work out of the box with Org[7]. Some MathJax display options can be configured via `org-html-mathjax-options`, or in the buffer. For example, with the following settings,

```
#+HTML_MATHJAX: align: left indent: 5em tagside: left font: Neo-Euler
```

equation labels will be displayed on the left margin and equations will be five ems from the left margin.

See the docstring of `org-html-mathjax-options` for all supported variables. The MathJax template can be configure via `org-html-mathjax-template`.

If you prefer, you can also request that LaTeX fragments are processed into small images that will be inserted into the browser page. Before the availability of MathJax, this was the default method for Org files. This method requires that the `dvipng` program, `dvisvgm` or `imagemagick` suite is available on your system. You can still get this processing with

```
#+OPTIONS: tex:dvipng
```

```
#+OPTIONS: tex:dvisvgm
```

or:

```
#+OPTIONS: tex:imagemagick
```

12.9.10 Text areas in HTML export

Before Org mode's Babel, one popular approach to publishing code in HTML was by using `:textarea`. The advantage of this approach was that copying and pasting was built into browsers with simple JavaScript commands. Even editing before pasting was made simple.

The HTML export back-end can create such text areas. It requires an `#+ATTR_HTML:` line as shown in the example below with the `:textarea` option. This must be followed by either an `example` or a `src` code block. Other Org block types will not honor the `:textarea` option.

By default, the HTML export back-end creates a text area 80 characters wide and height just enough to fit the content. Override these defaults with `:width` and `:height` options on the `#+ATTR_HTML:` line.

```
#+ATTR_HTML: :textarea t :width 40
#+BEGIN_EXAMPLE
  (defun org-xor (a b)
     "Exclusive or."
     (if a (not b) b))
#+END_EXAMPLE
```

[7] By default Org loads MathJax from cdnjs.com as recommended by MathJax.

12.9.11 CSS support

You can modify the CSS style definitions for the exported file. The HTML exporter assigns the following special CSS classes[8] to appropriate parts of the document—your style specifications may change these, in addition to any of the standard classes like for headlines, tables, etc.

`p.author`	author information, including email
`p.date`	publishing date
`p.creator`	creator info, about org mode version
`.title`	document title
`.subtitle`	document subtitle
`.todo`	TODO keywords, all not-done states
`.done`	the DONE keywords, all states that count as done
`.WAITING`	each TODO keyword also uses a class named after itself
`.timestamp`	timestamp
`.timestamp-kwd`	keyword associated with a timestamp, like SCHEDULED
`.timestamp-wrapper`	span around keyword plus timestamp
`.tag`	tag in a headline
`._HOME`	each tag uses itself as a class, "@" replaced by "_"
`.target`	target for links
`.linenr`	the line number in a code example
`.code-highlighted`	for highlighting referenced code lines
`div.outline-N`	div for outline level N (headline plus text))
`div.outline-text-N`	extra div for text at outline level N
`.section-number-N`	section number in headlines, different for each level
`.figure-number`	label like "Figure 1:"
`.table-number`	label like "Table 1:"
`.listing-number`	label like "Listing 1:"
`div.figure`	how to format an in-lined image
`pre.src`	formatted source code
`pre.example`	normal example
`p.verse`	verse paragraph
`div.footnotes`	footnote section headline
`p.footnote`	footnote definition paragraph, containing a footnote
`.footref`	a footnote reference number (always a <sup>)
`.footnum`	footnote number in footnote definition (always <sup>)
`.org-svg`	default class for a linked `.svg` image

The HTML export back-end includes a compact default style in each exported HTML file. To override the default style with another style, use these keywords in the Org file. They will replace the global defaults the HTML exporter uses.

```
#+HTML_HEAD: <link rel="stylesheet" type="text/css" href="style1.css" />
#+HTML_HEAD_EXTRA: <link rel="alternate stylesheet" type="text/css" href="st
```

To just turn off the default style, customize `org-html-head-include-default-style` variable, or use this option line in the Org file.

[8] If the classes on TODO keywords and tags lead to conflicts, use the variables `org-html-todo-kwd-class-prefix` and `org-html-tag-class-prefix` to make them unique.

```
#+OPTIONS: html-style:nil
```

For longer style definitions, either use several `#+HTML_HEAD` and `#+HTML_HEAD_EXTRA` lines, or use `<style> </style>` blocks around them. Both of these approaches can avoid referring to an external file.

In order to add styles to a sub-tree, use the `:HTML_CONTAINER_CLASS:` property to assign a class to the tree. In order to specify CSS styles for a particular headline, you can use the id specified in a `:CUSTOM_ID:` property.

Never change the `org-html-style-default` constant. Instead use other simpler ways of customizing as described above.

12.9.12 JavaScript supported display of web pages

Sebastian Rose has written a JavaScript program especially designed to enhance the web viewing experience of HTML files created with Org. This program enhances large files in two different ways of viewing. One is an *Info*-like mode where each section is displayed separately and navigation can be done with the *n* and *p* keys (and some other keys as well, press *?* for an overview of the available keys). The second one has a *folding* view, much like Org provides inside Emacs. The script is available at http://orgmode.org/org-info.js and the documentation at http://orgmode.org/worg/code/org-info-js/. The script is hosted on http://orgmode.org, but for reliability, prefer installing it on your own web server.

To use this program, just add this line to the Org file:

```
#+INFOJS_OPT: view:info toc:nil
```

The HTML header now has the code needed to automatically invoke the script. For setting options, use the syntax from the above line for options described below:

`path:` The path to the script. The default grabs the script from http://orgmode.org/org-info.js, but you might want to have a local copy and use a path like '`../scripts/org-info.js`'.

`view:` Initial view when the website is first shown. Possible values are:

`info`	Info-like interface with one section per page.
`overview`	Folding interface, initially showing only top-level.
`content`	Folding interface, starting with all headlines visible.
`showall`	Folding interface, all headlines and text visible.

`sdepth:` Maximum headline level that will still become an independent section for info and folding modes. The default is taken from `org-export-headline-levels` (= the `H` switch in `#+OPTIONS`). If this is smaller than in `org-export-headline-levels`, each info/folding section can still contain child headlines.

`toc:` Should the table of contents *initially* be visible? Even when `nil`, you can always get to the "toc" with *i*.

`tdepth:` The depth of the table of contents. The defaults are taken from the variables `org-export-headline-levels` and `org-export-with-toc`.

`ftoc:` Does the CSS of the page specify a fixed position for the "toc"? If yes, the toc will never be displayed as a section.

`ltoc:` Should there be short contents (children) in each section? Make this `above` if the section should be above initial text.

mouse: Headings are highlighted when the mouse is over them. Should be 'underline' (default) or a background color like '#cccccc'.

buttons: Should view-toggle buttons be everywhere? When nil (the default), only one such button will be present.

You can choose default values for these options by customizing the variable org-html-infojs-options. If you want the script to always apply to your pages, configure the variable org-html-use-infojs.

12.10 LaTeX export

The LaTeX export back-end can handle complex documents, incorporate standard or custom LaTeX document classes, generate documents using alternate LaTeX engines, and produce fully linked PDF files with indexes, bibliographies, and tables of contents, destined for interactive online viewing or high-quality print publication.

While the details are covered in-depth in this section, here are some quick references to variables for the impatient: for engines, see org-latex-compiler; for build sequences, see org-latex-pdf-process; for packages, see org-latex-default-packages-alist and org-latex-packages-alist.

An important note about the LaTeX export back-end: it is sensitive to blank lines in the Org document. That's because LaTeX itself depends on blank lines to tell apart syntactical elements, such as paragraphs.

12.10.1 LaTeX export commands

C-c C-e l l org-latex-export-to-latex
: Export as LaTeX file with a .tex extension. For myfile.org, Org exports to myfile.tex, overwriting without warning. *C-c C-e l l* Exports to LaTeX file.

C-c C-e l L org-latex-export-as-latex
: Export to a temporary buffer. Do not create a file.

C-c C-e l p org-latex-export-to-pdf
: Export as LaTeX file and convert it to PDF file.

C-c C-e l o
: Export as LaTeX file and convert it to PDF, then open the PDF using the default viewer.

The LaTeX export back-end can use any of these LaTeX engines: 'pdflatex', 'xelatex', and 'lualatex'. These engines compile LaTeX files with different compilers, packages, and output options. The LaTeX export back-end finds the compiler version to use from org-latex-compiler variable or the #+LATEX_COMPILER keyword in the Org file. See the docstring for the org-latex-default-packages-alist for loading packages with certain compilers. Also see org-latex-bibtex-compiler to set the bibliography compiler[9].

[9] This does not allow setting different bibliography compilers for different files. However, "smart" LaTeX compilation systems, such as 'latexmk', can select the correct bibliography compiler.

12.10.2 LaTeX specific export settings

The LaTeX export back-end has several additional keywords for customizing LaTeX output. Setting these keywords works similar to the general options (see Section 12.2 [Export settings], page 138).

'DESCRIPTION'
: The document's description. The description along with author name, keywords, and related file metadata are inserted in the output file by the 'hyperref' package. See `org-latex-hyperref-template` for customizing metadata items. See `org-latex-title-command` for typesetting description into the document's front matter. Use multiple `#+DESCRIPTION` lines for long descriptions.

'LATEX_CLASS'
: This is LaTeX document class, such as `article`, `report`, `book`, and so on, which contain predefined preamble and headline level mapping that the LaTeX export back-end needs. The back-end reads the default class name from the `org-latex-default-class` variable. Org has `article` as the default class. A valid default class must be an element of `org-latex-classes`.

'LATEX_CLASS_OPTIONS'
: Options the LaTeX export back-end uses when calling the LaTeX document class.

'LATEX_COMPILER'
: The compiler, such as 'pdflatex', 'xelatex', 'lualatex', for producing the PDF (`org-latex-compiler`).

'LATEX_HEADER'
: Arbitrary lines to add to the document's preamble, before the 'hyperref' settings. See `org-latex-classes` for adjusting the structure and order of the LaTeX headers.

'LATEX_HEADER_EXTRA'
: Arbitrary lines to add to the document's preamble, before the 'hyperref' settings. See `org-latex-classes` for adjusting the structure and order of the LaTeX headers.

'KEYWORDS'
: The keywords for the document. The description along with author name, keywords, and related file metadata are inserted in the output file by the 'hyperref' package. See `org-latex-hyperref-template` for customizing metadata items. See `org-latex-title-command` for typesetting description into the document's front matter. Use multiple `#+KEYWORDS` lines if necessary.

'SUBTITLE'
: The document's subtitle. It is typeset as per `org-latex-subtitle-format`. If `org-latex-subtitle-separate` is non-`nil`, it is typed as part of the '\title'-macro. See `org-latex-hyperref-template` for customizing metadata items. See `org-latex-title-command` for typesetting description into the document's front matter.

The following sections have further details.

12.10.3 LaTeX header and sectioning structure

The LaTeX export back-end converts the first three of Org's outline levels into LaTeX headlines. The remaining Org levels are exported as `itemize` or `enumerate` lists. To change this globally for the cut-off point between levels and lists, (see Section 12.2 [Export settings], page 138).

By default, the LaTeX export back-end uses the `article` class.

To change the default class globally, edit `org-latex-default-class`. To change the default class locally in an Org file, add option lines `#+LATEX_CLASS: myclass`. To change the default class for just a part of the Org file, set a sub-tree property, `EXPORT_LATEX_CLASS`. The class name entered here must be valid member of `org-latex-classes`. This variable defines a header template for each class into which the exporter splices the values of `org-latex-default-packages-alist` and `org-latex-packages-alist`. Use the same three variables to define custom sectioning or custom classes.

The LaTeX export back-end sends the `LATEX_CLASS_OPTIONS` keyword and `EXPORT_LATEX_CLASS_OPTIONS` property as options to the LaTeX `\documentclass` macro. The options and the syntax for specifying them, including enclosing them in square brackets, follow LaTeX conventions.

```
#+LATEX_CLASS_OPTIONS: [a4paper,11pt,twoside,twocolumn]
```

The LaTeX export back-end appends values from `LATEX_HEADER` and `LATEX_HEADER_EXTRA` keywords to the LaTeX header. The docstring for `org-latex-classes` explains in more detail. Also note that LaTeX export back-end does not append `LATEX_HEADER_EXTRA` to the header when previewing LaTeX snippets (see Section 11.8.2 [Previewing LaTeX fragments], page 135).

A sample Org file with the above headers:

```
#+LATEX_CLASS: article
#+LATEX_CLASS_OPTIONS: [a4paper]
#+LATEX_HEADER: \usepackage{xyz}

* Headline 1
  some text
* Headline 2
  some more text
```

12.10.4 Quoting LaTeX code

The LaTeX export back-end can insert any arbitrary LaTeX code, see Section 11.8 [Embedded LaTeX], page 134. There are three ways to embed such code in the Org file and they all use different quoting syntax.

Inserting in-line quoted with symbols:

```
Code embedded in-line @@latex:any arbitrary LaTeX code@@ in a paragraph.
```

Inserting as one or more keyword lines in the Org file:

```
#+LATEX: any arbitrary LaTeX code
```

Inserting as an export block in the Org file, where the back-end exports any code between begin and end markers:

```
#+BEGIN_EXPORT latex
any arbitrary LaTeX code
#+END_EXPORT
```

12.10.5 Tables in LaTeX export

The LaTeX export back-end can pass several LaTeX attributes for table contents and layout. Besides specifying label and caption (see Section 11.4 [Images and tables], page 131), the other valid LaTeX attributes include:

`:mode`
: The LaTeX export back-end wraps the table differently depending on the mode for accurate rendering of math symbols. Mode is either `table`, `math`, `inline-math` or `verbatim`. For `math` or `inline-math` mode, LaTeX export back-end wraps the table in a math environment, but every cell in it is exported as-is. The LaTeX export back-end determines the default mode from `org-latex-default-table-mode`. For , The LaTeX export back-end merges contiguous tables in the same mode into a single environment.

`:environment`
: Set the default LaTeX table environment for the LaTeX export back-end to use when exporting Org tables. Common LaTeX table environments are provided by these packages: `tabularx`, `longtable`, `array`, `tabu`, and `bmatrix`. For packages, such as `tabularx` and `tabu`, or any newer replacements, include them in the `org-latex-packages-alist` variable so the LaTeX export back-end can insert the appropriate load package headers in the converted LaTeX file. Look in the docstring for the `org-latex-packages-alist` variable for configuring these packages for LaTeX snippet previews, if any.

`:caption`
: Use `#+CAPTION` keyword to set a simple caption for a table (see Section 11.4 [Images and tables], page 131). For custom captions, use `:caption` attribute, which accepts raw LaTeX code. `:caption` value overrides `#+CAPTION` value.

`:float`
`:placement`
: The table environments by default are not floats in LaTeX. To make them floating objects use `:float` with one of the following options: `sideways`, `multicolumn`, `t`, and `nil`. Note that `sidewaystable` has been deprecated since Org 8.3. LaTeX floats can also have additional layout `:placement` attributes. These are the usual `[h t b p ! H]` permissions specified in square brackets. Note that for `:float sideways` tables, the LaTeX export back-end ignores `:placement` attributes.

`:align`
`:font`
`:width`
: The LaTeX export back-end uses these attributes for regular tables to set their alignments, fonts, and widths.

`:spread`
: When `:spread` is non-`nil`, the LaTeX export back-end spreads or shrinks the table by the `:width` for `tabu` and `longtabu` environments. `:spread` has no effect if `:width` is not set.

:booktabs
:center
:rmlines All three commands are toggles. :booktabs brings in modern typesetting enhancements to regular tables. The booktabs package has to be loaded through org-latex-packages-alist. :center is for centering the table. :rmlines removes all but the very first horizontal line made of ASCII characters from "table.el" tables only.

:math-prefix
:math-suffix
:math-arguments
The LaTeX export back-end inserts :math-prefix string value in a math environment before the table. The LaTeX export back-end inserts :math-suffix string value in a math environment after the table. The LaTeX export back-end inserts :math-arguments string value between the macro name and the table's contents. :math-arguments comes in use for matrix macros that require more than one argument, such as qbordermatrix.

LaTeX table attributes help formatting tables for a wide range of situations, such as matrix product or spanning multiple pages:

```
#+ATTR_LATEX: :environment longtable :align l|lp{3cm}r|l
| ..... | ..... |
| ..... | ..... |

#+ATTR_LATEX: :mode math :environment bmatrix :math-suffix \times
| a | b |
| c | d |
#+ATTR_LATEX: :mode math :environment bmatrix
| 1 | 2 |
| 3 | 4 |
```

Set the caption with the LaTeX command \bicaption{HeadingA}{HeadingB}:

```
#+ATTR_LATEX: :caption \bicaption{HeadingA}{HeadingB}
| ..... | ..... |
| ..... | ..... |
```

12.10.6 Images in LaTeX export

The LaTeX export back-end processes image links in Org files that do not have descriptions, such as these links '[[file:img.jpg]]' or '[[./img.jpg]]', as direct image insertions in the final PDF output. In the PDF, they are no longer links but actual images embedded on the page. The LaTeX export back-end uses \includegraphics macro to insert the image. But for TikZ[10] images, the back-end uses an \input macro wrapped within a tikzpicture environment.

For specifying image :width, :height, and other :options, use this syntax:

```
#+ATTR_LATEX: :width 5cm :options angle=90
[[./img/sed-hr4049.pdf]]
```

[10] http://sourceforge.net/projects/pgf/

For custom commands for captions, use the `:caption` attribute. It will override the default `#+CAPTION` value:

```
#+ATTR_LATEX: :caption \bicaption{HeadingA}{HeadingB}
[[./img/sed-hr4049.pdf]]
```

When captions follow the method as described in Section 11.4 [Images and tables], page 131, the LaTeX export back-end wraps the picture in a floating `figure` environment. To float an image without specifying a caption, set the `:float` attribute to one of the following:

- `t`: for a standard '`figure`' environment; used by default whenever an image has a caption.
- `multicolumn`: to span the image across multiple columns of a page; the back-end wraps the image in a `figure*` environment.
- `wrap`: for text to flow around the image on the right; the figure occupies the left half of the page.
- `sideways`: for a new page with the image sideways, rotated ninety degrees, in a `sidewaysfigure` environment; overrides `:placement` setting.
- `nil`: to avoid a `:float` even if using a caption.

Use the `placement` attribute to modify a floating environment's placement.

```
#+ATTR_LATEX: :float wrap :width 0.38\textwidth :placement
{r}{0.4\textwidth} [[./img/hst.png]]
```

The LaTeX export back-end centers all images by default. Setting `:center` attribute to `nil` disables centering. To disable centering globally, set `org-latex-images-centered` to `t`.

Set the `:comment-include` attribute to non-`nil` value for the LaTeX export back-end to comment out the `\includegraphics` macro.

12.10.7 Plain lists in LaTeX export

The LaTeX export back-end accepts the `:environment` and `:options` attributes for plain lists. Both attributes work together for customizing lists, as shown in the examples:

```
#+LATEX_HEADER: \usepackage[inline]{enumitem}
Some ways to say "Hello":
#+ATTR_LATEX: :environment itemize*
#+ATTR_LATEX: :options [label={}, itemjoin={,}, itemjoin*={, and}]
- Hola
- Bonjour
- Guten Tag.
```

Since LaTeX supports only four levels of nesting for lists, use an external package, such as '`enumitem`' in LaTeX, for levels deeper than four:

```
#+LATEX_HEADER: \usepackage{enumitem}
#+LATEX_HEADER: \renewlist{itemize}{itemize}{9}
#+LATEX_HEADER: \setlist[itemize]{label=$\circ$}
- One
  - Two
```

```
  - Three
    - Four
      - Five
```

12.10.8 Source blocks in LaTeX export

The LaTeX export back-end can make source code blocks into floating objects through the attributes `:float` and `:options`. For `:float`:

- `t`: makes a source block float; by default floats any source block with a caption.
- `multicolumn`: spans the source block across multiple columns of a page.
- `nil`: avoids a `:float` even if using a caption; useful for source code blocks that may not fit on a page.

```
#+ATTR_LATEX: :float nil
#+BEGIN_SRC emacs-lisp
Lisp code that may not fit in a single page.
#+END_SRC
```

The LaTeX export back-end passes string values in `:options` to LaTeX packages for customization of that specific source block. In the example below, the `:options` are set for Minted. Minted is a source code highlighting LaTeXpackage with many configurable options.

```
#+ATTR_LATEX: :options commentstyle=\bfseries
#+BEGIN_SRC emacs-lisp
  (defun Fib (n)
    (if (< n 2) n (+ (Fib (- n 1)) (Fib (- n 2)))))
#+END_SRC
```

To apply similar configuration options for all source blocks in a file, use the `org-latex-listings-options` and `org-latex-minted-options` variables.

12.10.9 Example blocks in LaTeX export

The LaTeX export back-end wraps the contents of example blocks in a '`verbatim`' environment. To change this behavior to use another environment globally, specify an appropriate export filter (see Section 12.17 [Advanced configuration], page 183). To change this behavior to use another environment for each block, use the `:environment` parameter to specify a custom environment.

```
#+ATTR_LATEX: :environment myverbatim
#+BEGIN_EXAMPLE
This sentence is false.
#+END_EXAMPLE
```

12.10.10 Special blocks in LaTeX export

For other special blocks in the Org file, the LaTeX export back-end makes a special environment of the same name. The back-end also takes `:options`, if any, and appends as-is to that environment's opening string. For example:

```
#+BEGIN_abstract
We demonstrate how to solve the Syracuse problem.
#+END_abstract
```

```
#+ATTR_LATEX: :options [Proof of important theorem]
#+BEGIN_proof
...
Therefore, any even number greater than 2 is the sum of two primes.
#+END_proof
```

exports to

```
\begin{abstract}
We demonstrate how to solve the Syracuse problem.
\end{abstract}

\begin{proof}[Proof of important theorem]
...
Therefore, any even number greater than 2 is the sum of two primes.
\end{proof}
```

If you need to insert a specific caption command, use `:caption` attribute. It will override standard `#+CAPTION` value, if any. For example:

```
#+ATTR_LATEX: :caption \MyCaption{HeadingA}
#+BEGIN_proof
...
#+END_proof
```

12.10.11 Horizontal rules in LaTeX export

The LaTeX export back-end converts horizontal rules by the specified `:width` and `:thickness` attributes. For example:

```
#+ATTR_LATEX: :width .6\textwidth :thickness 0.8pt
-----
```

12.11 Markdown export

The Markdown export back-end, `md`, converts an Org file to a Markdown format, as defined at `http://daringfireball.net/projects/markdown/`.

Since `md` is built on top of the HTML back-end, any Org constructs not supported by Markdown, such as tables, the underlying `html` back-end (see Section 12.9 [HTML export], page 149) converts them.

Markdown export commands

C-c C-e m m `org-md-export-to-markdown`
: Export to a text file with Markdown syntax. For `myfile.org`, Org exports to `myfile.md`, overwritten without warning.

C-c C-e m M `org-md-export-as-markdown`
: Export to a temporary buffer. Does not create a file.

C-c C-e m o
: Export as a text file with Markdown syntax, then open it.

Header and sectioning structure

Based on `org-md-headline-style`, markdown export can generate headlines of both `atx` and `setext` types. `atx` limits headline levels to two. `setext` limits headline levels to six. Beyond these limits, the export back-end converts headlines to lists. To set a limit to a level before the absolute limit (see Section 12.2 [Export settings], page 138).

12.12 OpenDocument Text export

The ODT export back-end handles creating of OpenDocument Text (ODT) format files. The format complies with *OpenDocument-v1.2 specification*[11] and is compatible with LibreOffice 3.4.

12.12.1 Pre-requisites for ODT export

The ODT export back-end relies on the `zip` program to create the final compressed ODT output. Check if `zip` is locally available and executable. Without `zip`, export cannot finish.

12.12.2 ODT export commands

C-c C-e o o `org-odt-export-to-odt`

Export as OpenDocument Text file.

If `org-odt-preferred-output-format` is specified, the ODT export back-end automatically converts the exported file to that format. See [Automatically exporting to other formats], page 166.

For `myfile.org`, Org exports to `myfile.odt`, overwriting without warning. The ODT export back-end exports a region only if a region was active. Note for exporting active regions, the `transient-mark-mode` has to be turned on.

If the selected region is a single tree, the ODT export back-end makes the tree head the document title. Incidentally, *C-c @* selects the current sub-tree. If the tree head entry has, or inherits, an `EXPORT_FILE_NAME` property, the ODT export back-end uses that for file name.

C-c C-e o O Export to an OpenDocument Text file format and open it.

When `org-odt-preferred-output-format` is specified, open the converted file instead. See [Automatically exporting to other formats], page 166.

12.12.3 ODT specific export settings

The ODT export back-end has several additional keywords for customizing ODT output. Setting these keywords works similar to the general options (see Section 12.2 [Export settings], page 138).

'`DESCRIPTION`'

This is the document's description, which the ODT export back-end inserts as document metadata. For long descriptions, use multiple `#+DESCRIPTION` lines.

'`KEYWORDS`'

The keywords for the document. The ODT export back-end inserts the description along with author name, keywords, and related file metadata as metadata in the output file. Use multiple `#+KEYWORDS` lines if necessary.

---

[11] Open Document Format for Office Applications (OpenDocument) Version 1.2

'ODT_STYLES_FILE'
: The ODT export back-end uses the `org-odt-styles-file` by default. See Section 12.12.5 [Applying custom styles], page 166 for details.

'SUBTITLE'
: The document subtitle.

12.12.4 Extending ODT export

The ODT export back-end can produce documents in other formats besides ODT using a specialized ODT converter process. Its common interface works with popular converters to produce formats such as 'doc', or convert a document from one format, say 'csv', to another format, say 'xls'.

Customize `org-odt-convert-process` variable to point to `unoconv`, which is the ODT's preferred converter. Working installations of LibreOffice would already have `unoconv` installed. Alternatively, other converters may be substituted here. See [Configuring a document converter], page 171.

Automatically exporting to other formats

If ODT format is just an intermediate step to get to other formats, such as 'doc', 'docx', 'rtf', or 'pdf', etc., then extend the ODT export back-end to directly produce that format. Specify the final format in the `org-odt-preferred-output-format` variable. This is one way to extend (see [Exporting to ODT], page 165).

Converting between document formats

The Org export back-end is made to be inter-operable with a wide range of text document format converters. Newer generation converters, such as LibreOffice and Pandoc, can handle hundreds of formats at once. Org provides a consistent interaction with whatever converter is installed. Here are some generic commands:

M-x org-odt-convert RET
: Convert an existing document from one format to another. With a prefix argument, opens the newly produced file.

12.12.5 Applying custom styles

The ODT export back-end comes with many OpenDocument styles (see [Working with OpenDocument style files], page 171). To expand or further customize these built-in style sheets, either edit the style sheets directly or generate them using an application such as LibreOffice. The example here shows creating a style using LibreOffice.

Applying custom styles: the easy way

1. Create a sample `example.org` file with settings as shown below, and export it to ODT format.

   ```
   #+OPTIONS: H:10 num:t
   ```

2. Open the above `example.odt` using LibreOffice. Use the `Stylist` to locate the target styles, which typically have the 'Org' prefix. Open one, modify, and save as either OpenDocument Text (`.odt`) or OpenDocument Template (`.ott`) file.

3. Customize the variable `org-odt-styles-file` and point it to the newly created file. For additional configuration options see [Overriding factory styles], page 172.

   To apply and ODT style to a particular file, use the `#+ODT_STYLES_FILE` option as shown in the example below:

   ```
   #+ODT_STYLES_FILE: "/path/to/example.ott"
   ```

   or

   ```
   #+ODT_STYLES_FILE: ("/path/to/file.ott" ("styles.xml" "image/hdr.png"))
   ```

Using third-party styles and templates

The ODT export back-end relies on many templates and style names. Using third-party styles and templates can lead to mismatches. Templates derived from built in ODT templates and styles seem to have fewer problems.

12.12.6 Links in ODT export

ODT export back-end creates native cross-references for internal links and Internet-style links for all other link types.

A link with no description and pointing to a regular—un-itemized—outline heading is replaced with a cross-reference and section number of the heading.

A '`\ref{label}`'-style reference to an image, table etc. is replaced with a cross-reference and sequence number of the labeled entity. See Section 12.12.10 [Labels and captions in ODT export], page 170.

12.12.7 Tables in ODT export

The ODT export back-end handles native Org mode tables (see Chapter 3 [Tables], page 19) and simple `table.el` tables. Complex `table.el` tables having column or row spans are not supported. Such tables are stripped from the exported document.

By default, the ODT export back-end exports a table with top and bottom frames and with ruled lines separating row and column groups (see Section 3.3 [Column groups], page 23). All tables are typeset to occupy the same width. The ODT export back-end honors any table alignments and relative widths for columns (see Section 3.2 [Column width and alignment], page 22).

Note that the ODT export back-end interprets column widths as weighted ratios, the default weight being 1.

Specifying `:rel-width` property on an `#+ATTR_ODT` line controls the width of the table. For example:

```
#+ATTR_ODT: :rel-width 50
| Area/Month    |   Jan |   Feb |   Mar |   Sum |
|---------------+-------+-------+-------+-------|
| /             |     < |       |       |     < |
| <l13>         |  <r5> |  <r5> |  <r5> |  <r6> |
| North America |     1 |    21 |   926 |   948 |
| Middle East   |     6 |    75 |   844 |   925 |
| Asia Pacific  |     9 |    27 |   790 |   826 |
|---------------+-------+-------+-------+-------|
```

```
| Sum                |      16 |     123 |    2560 |    2699 |
```

On export, the above table takes 50% of text width area. The exporter sizes the columns in the ratio: 13:5:5:5:6. The first column is left-aligned and rest of the columns, right-aligned. Vertical rules separate the header and the last column. Horizontal rules separate the header and the last row.

For even more customization, create custom table styles and associate them with a table using the #+ATTR_ODT line. See [Customizing tables in ODT export], page 173.

12.12.8 Images in ODT export

Embedding images

The ODT export back-end processes image links in Org files that do not have descriptions, such as these links '[[file:img.jpg]]' or '[[./img.jpg]]', as direct image insertions in the final output. Either of these examples works:

```
[[file:img.png]]
```

```
[[./img.png]]
```

Embedding clickable images

For clickable images, provide a link whose description is another link to an image file. For example, to embed a image org-mode-unicorn.png which when clicked jumps to http://Orgmode.org website, do the following

```
[[http://orgmode.org][./org-mode-unicorn.png]]
```

Sizing and scaling of embedded images

Control the size and scale of the embedded images with the #+ATTR_ODT attribute.

The ODT export back-end starts with establishing the size of the image in the final document. The dimensions of this size is measured in centimeters. The back-end then queries the image file for its dimensions measured in pixels. For this measurement, the back-end relies on ImageMagick's identify program or Emacs create-image and image-size API. ImageMagick is the preferred choice for large file sizes or frequent batch operations. The back-end then converts the pixel dimensions using org-odt-pixels-per-inch into the familiar 72 dpi or 96 dpi. The default value for this is in display-pixels-per-inch, which can be tweaked for better results based on the capabilities of the output device. Here are some common image scaling operations:

Explicitly size the image
: To embed img.png as a 10 cm x 10 cm image, do the following:

```
#+ATTR_ODT: :width 10 :height 10
[[./img.png]]
```

Scale the image
: To embed img.png at half its size, do the following:

```
#+ATTR_ODT: :scale 0.5
[[./img.png]]
```

Scale the image to a specific width
: To embed `img.png` with a width of 10 cm while retaining the original height:width ratio, do the following:

```
#+ATTR_ODT: :width 10
[[./img.png]]
```

Scale the image to a specific height
: To embed `img.png` with a height of 10 cm while retaining the original height:width ratio, do the following

```
#+ATTR_ODT: :height 10
[[./img.png]]
```

Anchoring of images

The ODT export back-end can anchor images to '`"as-char"`', '`"paragraph"`', or '`"page"`'. Set the preferred anchor using the `:anchor` property of the `#+ATTR_ODT` line.

To create an image that is anchored to a page:

```
#+ATTR_ODT: :anchor "page"
[[./img.png]]
```

12.12.9 Math formatting in ODT export

The ODT export back-end has special support built-in for handling math.

Working with LaTeX math snippets

LaTeX math snippets (see Section 11.8.1 [LaTeX fragments], page 134) can be embedded in an ODT document in one of the following ways:

1. MathML

   Add this line to the Org file. This option is activated on a per-file basis.

   ```
   #+OPTIONS: LaTeX:t
   ```

   With this option, LaTeX fragments are first converted into MathML fragments using an external LaTeX-to-MathML converter program. The resulting MathML fragments are then embedded as an OpenDocument Formula in the exported document.

   To specify the LaTeX-to-MathML converter, customize the variables `org-latex-to-mathml-convert-command` and `org-latex-to-mathml-jar-file`.

   To use MathToWeb[12] as the preferred converter, configure the above variables as

   ```
   (setq org-latex-to-mathml-convert-command
         "java -jar %j -unicode -force -df %o %I"
         org-latex-to-mathml-jar-file
         "/path/to/mathtoweb.jar")
   ```

   To use LaTeXML[13] use

   ```
   (setq org-latex-to-mathml-convert-command
         "latexmlmath \"%i\" --presentationmathml=%o")
   ```

[12] See MathToWeb.

[13] See `http://dlmf.nist.gov/LaTeXML/`.

To quickly verify the reliability of the LaTeX-to-MathML converter, use the following commands:

M-x org-odt-export-as-odf RET
: Convert a LaTeX math snippet to an OpenDocument formula (`.odf`) file.

M-x org-odt-export-as-odf-and-open RET
: Convert a LaTeX math snippet to an OpenDocument formula (`.odf`) file and open the formula file with the system-registered application.

2. PNG images

   Add this line to the Org file. This option is activated on a per-file basis.

   ```
   #+OPTIONS: tex:dvipng
   ```

   ```
   #+OPTIONS: tex:dvisvgm
   ```

   or:

   ```
   #+OPTIONS: tex:imagemagick
   ```

   Under this option, LaTeX fragments are processed into PNG or SVG images and the resulting images are embedded in the exported document. This method requires `dvipng` program, `dvisvgm` or `imagemagick` programs.

Working with MathML or OpenDocument formula files

When embedding LaTeX math snippets in ODT documents is not reliable, there is one more option to try. Embed an equation by linking to its MathML (`.mml`) source or its OpenDocument formula (`.odf`) file as shown below:

```
[[./equation.mml]]
```

or

```
[[./equation.odf]]
```

12.12.10 Labels and captions in ODT export

ODT format handles labeling and captioning of objects based on their types. Inline images, tables, LaTeX fragments, and Math formulas are numbered and captioned separately. Each object also gets a unique sequence number based on its order of first appearance in the Org file. Each category has its own sequence. A caption is just a label applied to these objects.

```
#+CAPTION: Bell curve
#+LABEL:   fig:SED-HR4049
[[./img/a.png]]
```

When rendered, it may show as follows in the exported document:

```
Figure 2: Bell curve
```

To modify the category component of the caption, customize the option `org-odt-category-map-alist`. For example, to tag embedded images with the string '`Illustration`' instead of the default string '`Figure`', use the following setting:

```
(setq org-odt-category-map-alist
      '(("__Figure__" "Illustration" "value" "Figure" org-odt--enumerable-image-p)))
```

With the above modification, the previous example changes to:

```
Illustration 2: Bell curve
```

12.12.11 Literal examples in ODT export

The ODT export back-end supports literal examples (see Section 11.5 [Literal examples], page 131) with full fontification. Internally, the ODT export back-end relies on `htmlfontify.el` to generate the style definitions needed for fancy listings. The auto-generated styles get '`OrgSrc`' prefix and inherit colors from the faces used by Emacs `font-lock` library for that source language.

For custom fontification styles, customize the `org-odt-create-custom-styles-for-srcblocks` option.

To turn off fontification of literal examples, customize the `org-odt-fontify-srcblocks` option.

12.12.12 Advanced topics in ODT export

The ODT export back-end has extensive features useful for power users and frequent uses of ODT formats.

Configuring a document converter

The ODT export back-end works with popular converters with little or no extra configuration. See Section 12.12.4 [Extending ODT export], page 166. The following is for unsupported converters or tweaking existing defaults.

1. Register the converter

   Add the name of the converter to the `org-odt-convert-processes` variable. Note that it also requires how the converter is invoked on the command line. See the variable's docstring for details.

2. Configure its capabilities

   Specify which formats the converter can handle by customizing the variable `org-odt-convert-capabilities`. Use the entry for the default values in this variable for configuring the new converter. Also see its docstring for details.

3. Choose the converter

   Select the newly added converter as the preferred one by customizing the option `org-odt-convert-process`.

Working with OpenDocument style files

This section explores the internals of the ODT exporter; the means by which it produces styled documents; the use of automatic and custom OpenDocument styles.

a) Factory styles

The ODT exporter relies on two files for generating its output. These files are bundled with the distribution under the directory pointed to by the variable `org-odt-styles-dir`. The two files are:

- `OrgOdtStyles.xml`

  This file contributes to the `styles.xml` file of the final '`ODT`' document. This file gets modified for the following purposes:

  1. To control outline numbering based on user settings.

2. To add styles generated by `htmlfontify.el` for fontification of code blocks.

- `OrgOdtContentTemplate.xml`

  This file contributes to the `content.xml` file of the final 'ODT' document. The contents of the Org outline are inserted between the '<office:text>'...'</office:text>' elements of this file.

  Apart from serving as a template file for the final `content.xml`, the file serves the following purposes:

  1. It contains automatic styles for formatting of tables which are referenced by the exporter.
  2. It contains '<text:sequence-decl>'...'</text:sequence-decl>' elements that control numbering of tables, images, equations, and similar entities.

b) Overriding factory styles

The following two variables control the location from where the ODT exporter picks up the custom styles and content template files. Customize these variables to override the factory styles used by the exporter.

- `org-odt-styles-file`

  The ODT export back-end uses the file pointed to by this variable, such as `styles.xml`, for the final output. It can take one of the following values:

  1. A `styles.xml` file

     Use this file instead of the default `styles.xml`

  2. A `.odt` or `.ott` file

     Use the `styles.xml` contained in the specified OpenDocument Text or Template file

  3. A `.odt` or `.ott` file and a subset of files contained within them

     Use the `styles.xml` contained in the specified OpenDocument Text or Template file. Additionally extract the specified member files and embed those within the final 'ODT' document.

     Use this option if the `styles.xml` file references additional files like header and footer images.

  4. `nil`

     Use the default `styles.xml`

- `org-odt-content-template-file`

  Use this variable to specify the blank `content.xml` that will be used in the final output.

Creating one-off styles

The ODT export back-end can read embedded raw OpenDocument XML from the Org file. Such direct formatting are useful for one-off instances.

1. Embedding ODT tags as part of regular text

   Enclose OpenDocument syntax in '`@@odt:...@@`' for inline markup. For example, to highlight a region of text do the following:

```
@@odt:<text:span text:style-name="Highlight">This is highlighted
text</text:span>@@. But this is regular text.
```

Hint: To see the above example in action, edit the `styles.xml` (see [Factory styles], page 171) and add a custom 'Highlight' style as shown below:

```
<style:style style:name="Highlight" style:family="text">
  <style:text-properties fo:background-color="#ff0000"/>
</style:style>
```

2. Embedding a one-line OpenDocument XML

   The ODT export back-end can read one-liner options with `#+ODT:` in the Org file. For example, to force a page break:

```
#+ODT: <text:p text:style-name="PageBreak"/>
```

Hint: To see the above example in action, edit your `styles.xml` (see [Factory styles], page 171) and add a custom 'PageBreak' style as shown below.

```
<style:style style:name="PageBreak" style:family="paragraph"
             style:parent-style-name="Text_20_body">
  <style:paragraph-properties fo:break-before="page"/>
</style:style>
```

3. Embedding a block of OpenDocument XML

   The ODT export back-end can also read ODT export blocks for OpenDocument XML. Such blocks use the `#+BEGIN_EXPORT odt`...`#+END_EXPORT` constructs.

   For example, to create a one-off paragraph that uses bold text, do the following:

```
#+BEGIN_EXPORT odt
<text:p text:style-name="Text_20_body_20_bold">
This paragraph is specially formatted and uses bold text.
</text:p>
#+END_EXPORT
```

Customizing tables in ODT export

Override the default table format by specifying a custom table style with the `#+ATTR_ODT` line. For a discussion on default formatting of tables see Section 12.12.7 [Tables in ODT export], page 167.

This feature closely mimics the way table templates are defined in the OpenDocument-v1.2 specification.[14]

For quick preview of this feature, install the settings below and export the table that follows:

```
(setq org-odt-table-styles
      (append org-odt-table-styles
              '(("TableWithHeaderRowAndColumn" "Custom"
                 ((use-first-row-styles . t)
                  (use-first-column-styles . t)))
                ("TableWithFirstRowandLastRow" "Custom"
                 ((use-first-row-styles . t)
                  (use-last-row-styles . t))))))
```

[14] OpenDocument-v1.2 Specification

```
#+ATTR_ODT: :style TableWithHeaderRowAndColumn
| Name  | Phone | Age |
| Peter |  1234 |  17 |
| Anna  |  4321 |  25 |
```

The example above used 'Custom' template and installed two table styles 'TableWithHeaderRowAndColumn' and 'TableWithFirstRowandLastRow'. **Important:** The OpenDocument styles needed for producing the above template were pre-defined. They are available in the section marked 'Custom Table Template' in OrgOdtContentTemplate.xml (see [Factory styles], page 172. For adding new templates, define new styles here.

To use this feature proceed as follows:

1. Create a table template[15]

   A table template is set of 'table-cell' and 'paragraph' styles for each of the following table cell categories:

   - Body
   - First column
   - Last column
   - First row
   - Last row
   - Even row
   - Odd row
   - Even column
   - Odd Column

   The names for the above styles must be chosen based on the name of the table template using a well-defined convention.

   The naming convention is better illustrated with an example. For a table template with the name 'Custom', the needed style names are listed in the following table.

Table cell type	table-cell style	paragraph style
Body	'CustomTableCell'	'CustomTableParagraph'
First column	'CustomFirstColumnTableCell'	'CustomFirstColumnTableParagraph'
Last column	'CustomLastColumnTableCell'	'CustomLastColumnTableParagraph'
First row	'CustomFirstRowTableCell'	'CustomFirstRowTableParagraph'
Last row	'CustomLastRowTableCell'	'CustomLastRowTableParagraph'
Even row	'CustomEvenRowTableCell'	'CustomEvenRowTableParagraph'
Odd row	'CustomOddRowTableCell'	'CustomOddRowTableParagraph'
Even column	'CustomEvenColumnTableCell'	'CustomEvenColumnTableParagraph'
Odd column	'CustomOddColumnTableCell'	'CustomOddColumnTableParagraph'

   To create a table template with the name 'Custom', define the above styles in the <office:automatic-styles>...</office:automatic-styles> element of the content template file (see [Factory styles], page 172).

[15] See the <table:table-template> element of the OpenDocument-v1.2 specification

2. Define a table style[16]

   To define a table style, create an entry for the style in the variable `org-odt-table-styles` and specify the following:
   - the name of the table template created in step (1)
   - the set of cell styles in that template that are to be activated

   For example, the entry below defines two different table styles '`TableWithHeaderRowAndColumn`' and '`TableWithFirstRowandLastRow`' based on the same template '`Custom`'. The styles achieve their intended effect by selectively activating the individual cell styles in that template.

```
(setq org-odt-table-styles
      (append org-odt-table-styles
              '(("TableWithHeaderRowAndColumn" "Custom"
                 ((use-first-row-styles . t)
                  (use-first-column-styles . t)))
                ("TableWithFirstRowandLastRow" "Custom"
                 ((use-first-row-styles . t)
                  (use-last-row-styles . t))))))
```

3. Associate a table with the table style

   To do this, specify the table style created in step (2) as part of the `ATTR_ODT` line as shown below.

```
#+ATTR_ODT: :style "TableWithHeaderRowAndColumn"
| Name  | Phone | Age |
| Peter |  1234 |  17 |
| Anna  |  4321 |  25 |
```

Validating OpenDocument XML

Sometimes ODT format files may not open due to `.odt` file corruption. To verify if the `.odt` file is corrupt, validate it against the OpenDocument RELAX NG Compact Syntax—RNC—schema. But first the `.odt` files have to be decompressed using '`zip`'. Note that `.odt` files are '`zip`' archives: See Info file `emacs`, node '`File Archives`'. The contents of `.odt` files are in `.xml`. For general help with validation—and schema-sensitive editing—of XML files: See Info file `nxml-mode`, node '`Introduction`'.

Customize `org-odt-schema-dir` to point to a directory with OpenDocument `.rnc` files and the needed schema-locating rules. The ODT export back-end takes care of updating the `rng-schema-locating-files`.

12.13 Org export

`org` export back-end creates a normalized version of the Org document in current buffer. The exporter evaluates Babel code (see Section 14.5 [Evaluating code blocks], page 203) and removes content specific to other back-ends.

[16] See the attributes `table:template-name`, `table:use-first-row-styles`, `table:use-last-row-styles`, `table:use-first-column-styles`, `table:use-last-column-styles`, `table:use-banding-rows-styles`, and `table:use-banding-column-styles` of the `<table:table>` element in the OpenDocument-v1.2 specification

Org export commands

C-c C-e O o `org-org-export-to-org`
: Export as an Org file with a `.org` extension. For `myfile.org`, Org exports to `myfile.org.org`, overwriting without warning.

C-c C-e O O `org-org-export-as-org`
: Export to a temporary buffer. Does not create a file.

C-c C-e O v
: Export to an Org file, then open it.

12.14 Texinfo export

The '`texinfo`' export back-end generates documents with Texinfo code that can compile to Info format.

12.14.1 Texinfo export commands

C-c C-e i t `org-texinfo-export-to-texinfo`
: Export as a Texinfo file with `.texi` extension. For `myfile.org`, Org exports to `myfile.texi`, overwriting without warning.

C-c C-e i i `org-texinfo-export-to-info`
: Export to Texinfo format first and then process it to make an Info file. To generate other formats, such as DocBook, customize the `org-texinfo-info-process` variable.

12.14.2 Texinfo specific export settings

The Texinfo export back-end has several additional keywords for customizing Texinfo output. Setting these keywords works similar to the general options (see Section 12.2 [Export settings], page 138).

'`SUBTITLE`'
: The document subtitle.

'`SUBAUTHOR`'
: The document subauthor.

'`TEXINFO_FILENAME`'
: The Texinfo filename.

'`TEXINFO_CLASS`'
: The default document class (`org-texinfo-default-class`), which must be a member of `org-texinfo-classes`.

'`TEXINFO_HEADER`'
: Arbitrary lines inserted at the end of the header.

'`TEXINFO_POST_HEADER`'
: Arbitrary lines inserted after the end of the header.

'`TEXINFO_DIR_CATEGORY`'
: The directory category of the document.

'TEXINFO_DIR_TITLE'
: The directory title of the document.

'TEXINFO_DIR_DESC'
: The directory description of the document.

'TEXINFO_PRINTED_TITLE'
: The printed title of the document.

12.14.3 Texinfo file header

After creating the header for a Texinfo file, the Texinfo back-end automatically generates a name and destination path for the Info file. To override this default with a more sensible path and name, specify the `#+TEXINFO_FILENAME` keyword.

Along with the output's file name, the Texinfo header also contains language details (see Section 12.2 [Export settings], page 138) and encoding system as set in the `org-texinfo-coding-system` variable. Insert `#+TEXINFO_HEADER` keywords for each additional command in the header, for example: @code{@synindex}.

Instead of repeatedly installing the same set of commands, define a class in `org-texinfo-classes` once, and then activate it in the document by setting the `#+TEXINFO_CLASS` keyword to that class.

12.14.4 Texinfo title and copyright page

The default template for hard copy output has a title page with `#+TITLE` and `#+AUTHOR` (see Section 12.2 [Export settings], page 138). To replace the regular `#+TITLE` with something different for the printed version, use the `#+TEXINFO_PRINTED_TITLE` and `#+SUBTITLE` keywords. Both expect raw Texinfo code for setting their values.

If one `#+AUTHOR` is not sufficient, add multiple `#+SUBAUTHOR` keywords. They have to be set in raw Texinfo code.

```
#+AUTHOR: Jane Smith
#+SUBAUTHOR: John Doe
#+TEXINFO_PRINTED_TITLE: This Long Title@inlinefmt{tex,@*} Is Broken in @TeX
```

Copying material is defined in a dedicated headline with a non-`nil` `:COPYING:` property. The back-end inserts the contents within a `@copying` command at the beginning of the document. The heading itself does not appear in the structure of the document.

Copyright information is printed on the back of the title page.

```
* Legalese
  :PROPERTIES:
  :COPYING: t
  :END:

  This is a short example of a complete Texinfo file, version 1.0.

  Copyright \copy 2016 Free Software Foundation, Inc.
```

12.14.5 Info directory file

The end result of the Texinfo export process is the creation of an Info file. This Info file's metadata has variables for category, title, and description: `#+TEXINFO_DIR_CATEGORY`,

#+TEXINFO_DIR_TITLE, and #+TEXINFO_DIR_DESC that establish where in the Info hierarchy the file fits.

Here is an example that writes to the Info directory file:

```
#+TEXINFO_DIR_CATEGORY: Emacs
#+TEXINFO_DIR_TITLE: Org Mode: (org)
#+TEXINFO_DIR_DESC: Outline-based notes management and organizer
```

12.14.6 Headings and sectioning structure

The Texinfo export back-end uses a pre-defined scheme to convert Org headlines to an equivalent Texinfo structuring commands. A scheme like this maps top-level headlines to numbered chapters tagged as @chapter and lower-level headlines to unnumbered chapters tagged as @unnumbered. To override such mappings to introduce @part or other Texinfo structuring commands, define a new class in org-texinfo-classes. Activate the new class with the #+TEXINFO_CLASS keyword. When no new class is defined and activated, the Texinfo export back-end defaults to the org-texinfo-default-class.

If an Org headline's level has no associated Texinfo structuring command, or is below a certain threshold (see Section 12.2 [Export settings], page 138), then the Texinfo export back-end makes it into a list item.

The Texinfo export back-end makes any headline with a non-nil :APPENDIX: property into an appendix. This happens independent of the Org headline level or the #+TEXINFO_CLASS.

The Texinfo export back-end creates a menu entry after the Org headline for each regular sectioning structure. To override this with a shorter menu entry, use the :ALT_TITLE: property (see Section 12.3 [Table of contents], page 141). Texinfo menu entries also have an option for a longer :DESCRIPTION: property. Here's an example that uses both to override the default menu entry:

```
* Controlling Screen Display
  :PROPERTIES:
  :ALT_TITLE: Display
  :DESCRIPTION: Controlling Screen Display
  :END:
```

The text before the first headline belongs to the 'Top' node, i.e., the node in which a reader enters an Info manual. As such, it is expected not to appear in printed output generated from the .texi file. See Info file texinfo, node 'The Top Node', for more information.

12.14.7 Indices

The Texinfo export back-end recognizes these indexing keywords if used in the Org file: #+CINDEX, #+FINDEX, #+KINDEX, #+PINDEX, #+TINDEX, and #+VINDEX. Write their value as verbatim Texinfo code; in particular, '{', '}' and '@' characters need to be escaped with '@' if they not belong to a Texinfo command.

```
#+CINDEX: Defining indexing entries
```

For the back-end to generate an index entry for a headline, set the :INDEX: property to 'cp' or 'vr'. These abbreviations come from Texinfo that stand for concept index and variable index. The Texinfo manual has abbreviations for all other kinds of indexes. The back-end exports the headline as an unnumbered chapter or section command, and then inserts the index after its contents.

```
* Concept Index
  :PROPERTIES:
  :INDEX: cp
  :END:
```

12.14.8 Quoting Texinfo code

Use any of the following three methods to insert or escape raw Texinfo code:

```
Richard @@texinfo:@sc{@@Stallman@@texinfo:}@@ commence' GNU.

#+TEXINFO: @need800
This paragraph is preceded by...

#+BEGIN_EXPORT texinfo
@auindex Johnson, Mark
@auindex Lakoff, George
#+END_EXPORT
```

12.14.9 Plain lists in Texinfo export

The Texinfo export back-end by default converts description lists in the Org file using the default command `@table`, which results in a table with two columns. To change this behavior, specify `:table-type` with `ftable` or `vtable` attributes. For more information, See Info file `texinfo`, node '`Two-column Tables`'.

The Texinfo export back-end by default also applies a text highlight based on the defaults stored in `org-texinfo-table-default-markup`. To override the default highlight command, specify another one with the `:indic` attribute.

Org syntax is limited to one entry per list item. Nevertheless, the Texinfo export back-end can split that entry according to any text provided through the `:sep` attribute. Each part then becomes a new entry in the first column of the table.

The following example illustrates all the attributes above:

```
#+ATTR_TEXINFO: :table-type vtable :sep , :indic asis
- foo, bar :: This is the common text for variables foo and bar.
```

becomes

```
@vtable @asis
@item foo
@itemx bar
This is the common text for variables foo and bar.
@end table
```

12.14.10 Tables in Texinfo export

When exporting tables, the Texinfo export back-end uses the widest cell width in each column. To override this and instead specify as fractions of line length, use the `:columns` attribute. See example below.

```
#+ATTR_TEXINFO: :columns .5 .5
| a cell | another cell |
```

12.14.11 Images in Texinfo export

Insert a file link to the image in the Org file, and the Texinfo export back-end inserts the image. These links must have the usual supported image extensions and no descriptions. To scale the image, use `:width` and `:height` attributes. For alternate text, use `:alt` and specify the text using Texinfo code, as shown in the example:

```
#+ATTR_TEXINFO: :width 1in :alt Alternate @i{text}
[[ridt.pdf]]
```

12.14.12 Special blocks

The Texinfo export back-end converts special blocks to commands with the same name. It also adds any `:options` attributes to the end of the command, as shown in this example:

```
#+ATTR_TEXINFO: :options org-org-export-to-org ...
#+begin_defun
A somewhat obsessive function.
#+end_defun
```

becomes

```
@defun org-org-export-to-org ...
A somewhat obsessive function.
@end defun
```

12.14.13 A Texinfo example

Here is a more detailed example Org file. See Section "GNU Sample Texts" in *GNU Texinfo Manual* for an equivalent example using Texinfo code.

```
#+TITLE: GNU Sample {{{version}}}
#+SUBTITLE: for version {{{version}}}, {{{updated}}}
#+AUTHOR: A.U. Thor
#+EMAIL: bug-sample@gnu.org

#+OPTIONS: ':t toc:t author:t email:t
#+LANGUAGE: en

#+MACRO: version 2.0
#+MACRO: updated last updated 4 March 2014

#+TEXINFO_FILENAME: sample.info
#+TEXINFO_HEADER: @syncodeindex pg cp

#+TEXINFO_DIR_CATEGORY: Texinfo documentation system
#+TEXINFO_DIR_TITLE: sample: (sample)
#+TEXINFO_DIR_DESC: Invoking sample

#+TEXINFO_PRINTED_TITLE: GNU Sample
```

```
This manual is for GNU Sample (version {{{version}}},
{{{updated}}}).

* Copying
  :PROPERTIES:
  :COPYING:  t
  :END:

  This manual is for GNU Sample (version {{{version}}},
  {{{updated}}}), which is an example in the Texinfo documentation.

  Copyright \copy 2016 Free Software Foundation, Inc.

  #+BEGIN_QUOTE
  Permission is granted to copy, distribute and/or modify this
  document under the terms of the GNU Free Documentation License,
  Version 1.3 or any later version published by the Free Software
  Foundation; with no Invariant Sections, with no Front-Cover Texts,
  and with no Back-Cover Texts.  A copy of the license is included in
  the section entitled "GNU Free Documentation License".
  #+END_QUOTE

* Invoking sample

  #+PINDEX: sample
  #+CINDEX: invoking @command{sample}

  This is a sample manual.  There is no sample program to invoke, but
  if there were, you could see its basic usage and command line
  options here.

* GNU Free Documentation License
  :PROPERTIES:
  :APPENDIX: t
  :END:

  #+TEXINFO: @include fdl.texi

* Index
  :PROPERTIES:
  :INDEX:    cp
  :END:
```

12.15 iCalendar export

A large part of Org mode's inter-operability success is its ability to easily export to or import from external applications. The iCalendar export back-end takes calendar data from Org files and exports to the standard iCalendar format.

The iCalendar export back-end can also incorporate TODO entries based on the configuration of the `org-icalendar-include-todo` variable. The back-end exports plain timestamps as VEVENT, TODO items as VTODO, and also create events from deadlines that are in non-TODO items. The back-end uses the deadlines and scheduling dates in Org TODO items for setting the start and due dates for the iCalendar TODO entry. Consult the `org-icalendar-use-deadline` and `org-icalendar-use-scheduled` variables for more details.

For tags on the headline, the iCalendar export back-end makes them into iCalendar categories. To tweak the inheritance of tags and TODO states, configure the variable `org-icalendar-categories`. To assign clock alarms based on time, configure the `org-icalendar-alarm-time` variable.

The iCalendar format standard requires globally unique identifier—UID—for each entry. The iCalendar export back-end creates UIDs during export. To save a copy of the UID in the Org file set the variable `org-icalendar-store-UID`. The back-end looks for the `:ID:` property of the entry for re-using the same UID for subsequent exports.

Since a single Org entry can result in multiple iCalendar entries—as timestamp, deadline, scheduled item, or TODO item—Org adds prefixes to the UID, depending on which part of the Org entry triggered the creation of the iCalendar entry. Prefixing ensures UIDs remains unique, yet enable synchronization programs trace the connections.

C-c C-e c f `org-icalendar-export-to-ics`
: Create iCalendar entries from the current Org buffer and store them in the same directory, using a file extension `.ics`.

C-c C-e c a `org-icalendar-export-agenda-files`
: Create iCalendar entries from Org files in `org-agenda-files` and store in a separate iCalendar file for each Org file.

C-c C-e c c `org-icalendar-combine-agenda-files`
: Create a combined iCalendar file from Org files in `org-agenda-files` and write it to `org-icalendar-combined-agenda-file` file name.

The iCalendar export back-end includes SUMMARY, DESCRIPTION, LOCATION and TIMEZONE properties from the Org entries when exporting. To force the back-end to inherit the LOCATION and TIMEZONE properties, configure the `org-use-property-inheritance` variable.

When Org entries do not have SUMMARY, DESCRIPTION and LOCATION properties, the iCalendar export back-end derives the summary from the headline, and derives the description from the body of the Org item. The `org-icalendar-include-body` variable limits the maximum number of characters of the content are turned into its description.

The TIMEZONE property can be used to specify a per-entry time zone, and will be applied to any entry with timestamp information. Time zones should be specified as per the IANA time zone database format, e.g. "Asia/Almaty". Alternately, the property value can be "UTC", to force UTC time for this entry only.

Exporting to iCalendar format depends in large part on the capabilities of the destination application. Some are more lenient than others. Consult the Org mode FAQ for advice on specific applications.

12.16 Other built-in back-ends

Other export back-ends included with Org are:

- `ox-man.el`: export to a man page.

To activate such back-ends, either customize `org-export-backends` or load directly with `(require 'ox-man)`. On successful load, the back-end adds new keys in the export dispatcher (see Section 12.1 [The export dispatcher], page 137).

Follow the comment section of such files, for example, `ox-man.el`, for usage and configuration details.

12.17 Advanced configuration

Hooks

The export process executes two hooks before the actual exporting begins. The first hook, `org-export-before-processing-hook`, runs before any expansions of macros, Babel code, and include keywords in the buffer. The second hook, `org-export-before-parsing-hook`, runs before the buffer is parsed. Both hooks are specified as functions, see example below. Their main use is for heavy duty structural modifications of the Org content. For example, removing every headline in the buffer during export:

```
(defun my-headline-removal (backend)
  "Remove all headlines in the current buffer.
BACKEND is the export back-end being used, as a symbol."
  (org-map-entries
   (lambda () (delete-region (point) (progn (forward-line) (point))))))

(add-hook 'org-export-before-parsing-hook 'my-headline-removal)
```

Note that the hook function must have a mandatory argument that is a symbol for the back-end.

Filters

The Org export process relies on filters to process specific parts of conversion process. Filters are just lists of functions to be applied to certain parts for a given back-end. The output from the first function in the filter is passed on to the next function in the filter. The final output is the output from the final function in the filter.

The Org export process has many filter sets applicable to different types of objects, plain text, parse trees, export options, and final output formats. The filters are named after the element type or object type: `org-export-filter-TYPE-functions`, where `TYPE` is the type targeted by the filter. Valid types are:

body	bold	babel-call
center-block	clock	code
diary-sexp	drawer	dynamic-block

entity
export-snippet
footnote-definition
horizontal-rule
inlinetask
keyword
line-break
options
plain-list
property-drawer
section
statistics-cookie
superscript
table-row
underline
example-block
final-output
footnote-reference
inline-babel-call
italic
latex-environment
link
paragraph
plain-text
quote-block
special-block
strike-through
table
target
verbatim
export-block
fixed-width
headline
inline-src-block
item
latex-fragment
node-property
parse-tree
planning
radio-target
src-block
subscript
table-cell
timestamp
verse-block

Here is an example filter that replaces non-breaking spaces ~ in the Org buffer with _ for the LaTeX back-end.

```
(defun my-latex-filter-nobreaks (text backend info)
  "Ensure \"_\" are properly handled in LaTeX export."
  (when (org-export-derived-backend-p backend 'latex)
        (replace-regexp-in-string "_" "~" text)))

(add-to-list 'org-export-filter-plain-text-functions
             'my-latex-filter-nobreaks)
```

A filter requires three arguments: the code to be transformed, the name of the back-end, and some optional information about the export process. The third argument can be safely ignored. Note the use of `org-export-derived-backend-p` predicate that tests for `latex` back-end or any other back-end, such as `beamer`, derived from `latex`.

Defining filters for individual files

The Org export can filter not just for back-ends, but also for specific files through the `#+BIND` keyword. Here is an example with two filters; one removes brackets from time stamps, and the other removes strike-through text. The filter functions are defined in a '`src`' code block in the same Org file, which is a handy location for debugging.

```
#+BIND: org-export-filter-timestamp-functions (tmp-f-timestamp)
#+BIND: org-export-filter-strike-through-functions (tmp-f-strike-through)
#+begin_src emacs-lisp :exports results :results none
  (defun tmp-f-timestamp (s backend info)
    (replace-regexp-in-string "&[lg]t;\\|[][]" "" s))
  (defun tmp-f-strike-through (s backend info) "")
#+end_src
```

Extending an existing back-end

Some parts of the conversion process can be extended for certain elements so as to introduce a new or revised translation. That is how the HTML export back-end was extended to handle Markdown format. The extensions work seamlessly so any aspect of filtering not done by the extended back-end is handled by the original back-end. Of all the export customization in Org, extending is very powerful as it operates at the parser level.

For this example, make the `ascii` back-end display the language used in a source code block. Also make it display only when some attribute is non-`nil`, like the following:

```
#+ATTR_ASCII: :language t
```

Then extend `ascii` back-end with a custom `my-ascii` back-end.

```
(defun my-ascii-src-block (src-block contents info)
  "Transcode a SRC-BLOCK element from Org to ASCII.
CONTENTS is nil.  INFO is a plist used as a communication
channel."
  (if (not (org-export-read-attribute :attr_ascii src-block :language))
    (org-export-with-backend 'ascii src-block contents info)
  (concat
   (format ",--[ %s ]--\n%s`----"
           (org-element-property :language src-block)
           (replace-regexp-in-string
            "^" "| "
            (org-element-normalize-string
             (org-export-format-code-default src-block info)))))))

(org-export-define-derived-backend 'my-ascii 'ascii
  :translate-alist '((src-block . my-ascii-src-block)))
```

The `my-ascii-src-block` function looks at the attribute above the current element. If not true, hands over to `ascii` back-end. If true, which it is in this example, it creates a box around the code and leaves room for the inserting a string for language. The last form creates the new back-end that springs to action only when translating `src-block` type elements.

To use the newly defined back-end, call the following from an Org buffer:

```
(org-export-to-buffer 'my-ascii "*Org MY-ASCII Export*")
```

Further steps to consider would be an interactive function, self-installing an item in the export dispatcher menu, and other user-friendly improvements.

12.18 Export in foreign buffers

The export back-ends in Org often include commands to convert selected regions. A convenient feature of this in-place conversion is that the exported output replaces the original source. Here are such functions:

`org-html-convert-region-to-html`
: Convert the selected region into HTML.

`org-latex-convert-region-to-latex`
: Convert the selected region into LaTeX.

`org-texinfo-convert-region-to-texinfo`
: Convert the selected region into `Texinfo`.

`org-md-convert-region-to-md`
: Convert the selected region into `MarkDown`.

In-place conversions are particularly handy for quick conversion of tables and lists in foreign buffers. For example, turn on the minor mode `M-x orgstruct-mode` in an HTML buffer, then use the convenient Org keyboard commands to create a list, select it, and covert it to HTML with `M-x org-html-convert-region-to-html RET`.

13 Publishing

Org includes a publishing management system that allows you to configure automatic HTML conversion of *projects* composed of interlinked org files. You can also configure Org to automatically upload your exported HTML pages and related attachments, such as images and source code files, to a web server.

You can also use Org to convert files into PDF, or even combine HTML and PDF conversion so that files are available in both formats on the server.

Publishing has been contributed to Org by David O'Toole.

13.1 Configuration

Publishing needs significant configuration to specify files, destination and many other properties of a project.

13.1.1 The variable `org-publish-project-alist`

Publishing is configured almost entirely through setting the value of one variable, called `org-publish-project-alist`. Each element of the list configures one project, and may be in one of the two following forms:

```
("project-name" :property value :property value ...)
```

i.e., a well-formed property list with alternating keys and values

or

```
("project-name" :components ("project-name" "project-name" ...))
```

In both cases, projects are configured by specifying property values. A project defines the set of files that will be published, as well as the publishing configuration to use when publishing those files. When a project takes the second form listed above, the individual members of the `:components` property are taken to be sub-projects, which group together files requiring different publishing options. When you publish such a "meta-project", all the components will also be published, in the sequence given.

13.1.2 Sources and destinations for files

Most properties are optional, but some should always be set. In particular, Org needs to know where to look for source files, and where to put published files.

`:base-directory` Directory containing publishing source files

`:publishing-directory` Directory where output files will be published. You can directly publish to a web server using a file name syntax appropriate for the Emacs `tramp` package. Or you can publish to a local directory and use external tools to upload your website (see Section 13.2 [Uploading files], page 195).

`:preparation-function` Function or list of functions to be called before starting the publishing process, for example, to run `make` for updating files to be published. Each preparation function is called with a single argument, the project property list.

`:completion-function` Function or list of functions called after finishing the publishing process, for example, to change permissions of the resulting files. Each completion function is called with a single argument, the project property list.

13.1.3 Selecting files

By default, all files with extension `.org` in the base directory are considered part of the project. This can be modified by setting the properties

`:base-extension` Extension (without the dot!) of source files. This actually is a regular expression. Set this to the symbol `any` if you want to get all files in `:base-directory`, even without extension.

`:exclude` Regular expression to match file names that should not be published, even though they have been selected on the basis of their extension.

`:include` List of files to be included regardless of `:base-extension` and `:exclude`.

`:recursive` non-`nil` means, check base-directory recursively for files to publish.

13.1.4 Publishing action

Publishing means that a file is copied to the destination directory and possibly transformed in the process. The default transformation is to export Org files as HTML files, and this is done by the function `org-html-publish-to-html`, which calls the HTML exporter (see Section 12.9 [HTML export], page 149). But you also can publish your content as PDF files using `org-latex-publish-to-pdf` or as `ascii`, `Texinfo`, etc., using the corresponding functions.

If you want to publish the Org file as an `.org` file but with the *archived*, *commented* and *tag-excluded* trees removed, use the function `org-org-publish-to-org`. This will produce `file.org` and put it in the publishing directory. If you want a htmlized version of this file, set the parameter `:htmlized-source` to `t`, it will produce `file.org.html` in the publishing directory[1].

Other files like images only need to be copied to the publishing destination. For this you can use `org-publish-attachment`. For non-org files, you always need to specify the publishing function:

`:publishing-function` Function executing the publication of a file. This may also be a list of functions, which will all be called in turn.

`:htmlized-source` non-`nil` means, publish htmlized source.

The function must accept three arguments: a property list containing at least a `:publishing-directory` property, the name of the file to be published and the path to

[1] If the publishing directory is the same than the source directory, `file.org` will be exported as `file.org.org`, so probably don't want to do this.

the publishing directory of the output file. It should take the specified file, make the necessary transformation (if any) and place the result into the destination folder.

13.1.5 Options for the exporters

The property list can be used to set export options during the publishing process. In most cases, these properties correspond to user variables in Org. While some properties are available for all export back-ends, most of them are back-end specific. The following sections list properties along with the variable they belong to. See the documentation string of these options for details.

When a property is given a value in `org-publish-project-alist`, its setting overrides the value of the corresponding user variable (if any) during publishing. Options set within a file (see Section 12.2 [Export settings], page 138), however, override everything.

Generic properties

`:archived-trees`	`org-export-with-archived-trees`
`:exclude-tags`	`org-export-exclude-tags`
`:headline-levels`	`org-export-headline-levels`
`:language`	`org-export-default-language`
`:preserve-breaks`	`org-export-preserve-breaks`
`:section-numbers`	`org-export-with-section-numbers`
`:select-tags`	`org-export-select-tags`
`:with-author`	`org-export-with-author`
`:with-broken-links`	`org-export-with-broken-links`
`:with-clocks`	`org-export-with-clocks`
`:with-creator`	`org-export-with-creator`
`:with-date`	`org-export-with-date`
`:with-drawers`	`org-export-with-drawers`
`:with-email`	`org-export-with-email`
`:with-emphasize`	`org-export-with-emphasize`
`:with-fixed-width`	`org-export-with-fixed-width`
`:with-footnotes`	`org-export-with-footnotes`
`:with-latex`	`org-export-with-latex`
`:with-planning`	`org-export-with-planning`
`:with-priority`	`org-export-with-priority`
`:with-properties`	`org-export-with-properties`
`:with-special-strings`	`org-export-with-special-strings`
`:with-sub-superscript`	`org-export-with-sub-superscripts`
`:with-tables`	`org-export-with-tables`
`:with-tags`	`org-export-with-tags`
`:with-tasks`	`org-export-with-tasks`
`:with-timestamps`	`org-export-with-timestamps`
`:with-title`	`org-export-with-title`
`:with-toc`	`org-export-with-toc`
`:with-todo-keywords`	`org-export-with-todo-keywords`

ASCII specific properties

:ascii-bullets	org-ascii-bullets
:ascii-caption-above	org-ascii-caption-above
:ascii-charset	org-ascii-charset
:ascii-global-margin	org-ascii-global-margin
:ascii-format-drawer-function	org-ascii-format-drawer-function
:ascii-format-inlinetask-function	org-ascii-format-inlinetask-function
:ascii-headline-spacing	org-ascii-headline-spacing
:ascii-indented-line-width	org-ascii-indented-line-width
:ascii-inlinetask-width	org-ascii-inlinetask-width
:ascii-inner-margin	org-ascii-inner-margin
:ascii-links-to-notes	org-ascii-links-to-notes
:ascii-list-margin	org-ascii-list-margin
:ascii-paragraph-spacing	org-ascii-paragraph-spacing
:ascii-quote-margin	org-ascii-quote-margin
:ascii-table-keep-all-vertical-lines	org-ascii-table-keep-all-vertical-lines
:ascii-table-use-ascii-art	org-ascii-table-use-ascii-art
:ascii-table-widen-columns	org-ascii-table-widen-columns
:ascii-text-width	org-ascii-text-width
:ascii-underline	org-ascii-underline
:ascii-verbatim-format	org-ascii-verbatim-format

Beamer specific properties

:beamer-theme	org-beamer-theme
:beamer-column-view-format	org-beamer-column-view-format
:beamer-environments-extra	org-beamer-environments-extra
:beamer-frame-default-options	org-beamer-frame-default-options
:beamer-outline-frame-options	org-beamer-outline-frame-options
:beamer-outline-frame-title	org-beamer-outline-frame-title
:beamer-subtitle-format	org-beamer-subtitle-format

HTML specific properties

:html-allow-name-attribute-in-anchors	org-html-allow-name-attribute-in-anchors
:html-checkbox-type	org-html-checkbox-type
:html-container	org-html-container-element
:html-divs	org-html-divs
:html-doctype	org-html-doctype
:html-extension	org-html-extension
:html-footnote-format	org-html-footnote-format
:html-footnote-separator	org-html-footnote-separator
:html-footnotes-section	org-html-footnotes-section
:html-format-drawer-function	org-html-format-drawer-function
:html-format-headline-function	org-html-format-headline-function
:html-format-inlinetask-function	org-html-format-inlinetask-function
:html-head-extra	org-html-head-extra
:html-head-include-default-style	org-html-head-include-default-style

:html-head-include-scripts	org-html-head-include-scripts
:html-head	org-html-head
:html-home/up-format	org-html-home/up-format
:html-html5-fancy	org-html-html5-fancy
:html-indent	org-html-indent
:html-infojs-options	org-html-infojs-options
:html-infojs-template	org-html-infojs-template
:html-inline-image-rules	org-html-inline-image-rules
:html-inline-images	org-html-inline-images
:html-link-home	org-html-link-home
:html-link-org-files-as-html	org-html-link-org-files-as-html
:html-link-up	org-html-link-up
:html-link-use-abs-url	org-html-link-use-abs-url
:html-mathjax-options	org-html-mathjax-options
:html-mathjax-template	org-html-mathjax-template
:html-metadata-timestamp-format	org-html-metadata-timestamp-format
:html-postamble-format	org-html-postamble-format
:html-postamble	org-html-postamble
:html-preamble-format	org-html-preamble-format
:html-preamble	org-html-preamble
:html-table-align-individual-fields	org-html-table-align-individual-fi
:html-table-attributes	org-html-table-default-attributes
:html-table-caption-above	org-html-table-caption-above
:html-table-data-tags	org-html-table-data-tags
:html-table-header-tags	org-html-table-header-tags
:html-table-row-tags	org-html-table-row-tags
:html-table-use-header-tags-for-first-column	org-html-table-use-header-tags-for
:html-tag-class-prefix	org-html-tag-class-prefix
:html-text-markup-alist	org-html-text-markup-alist
:html-todo-kwd-class-prefix	org-html-todo-kwd-class-prefix
:html-toplevel-hlevel	org-html-toplevel-hlevel
:html-use-infojs	org-html-use-infojs
:html-validation-link	org-html-validation-link
:html-viewport	org-html-viewport
:html-xml-declaration	org-html-xml-declaration

LaTeX specific properties

:latex-active-timestamp-format	org-latex-active-timestamp-format
:latex-caption-above	org-latex-caption-above
:latex-classes	org-latex-classes
:latex-class	org-latex-default-class
:latex-compiler	org-latex-compiler
:latex-default-figure-position	org-latex-default-figure-position
:latex-default-table-environment	org-latex-default-table-environment
:latex-default-table-mode	org-latex-default-table-mode
:latex-diary-timestamp-format	org-latex-diary-timestamp-format

:latex-footnote-defined-format	org-latex-footnote-defined-format
:latex-footnote-separator	org-latex-footnote-separator
:latex-format-drawer-function	org-latex-format-drawer-function
:latex-format-headline-function	org-latex-format-headline-function
:latex-format-inlinetask-function	org-latex-format-inlinetask-function
:latex-hyperref-template	org-latex-hyperref-template
:latex-image-default-height	org-latex-image-default-height
:latex-image-default-option	org-latex-image-default-option
:latex-image-default-width	org-latex-image-default-width
:latex-images-centered	org-latex-images-centered
:latex-inactive-timestamp-format	org-latex-inactive-timestamp-format
:latex-inline-image-rules	org-latex-inline-image-rules
:latex-link-with-unknown-path-format	org-latex-link-with-unknown-path-format
:latex-listings-langs	org-latex-listings-langs
:latex-listings-options	org-latex-listings-options
:latex-listings	org-latex-listings
:latex-minted-langs	org-latex-minted-langs
:latex-minted-options	org-latex-minted-options
:latex-prefer-user-labels	org-latex-prefer-user-labels
:latex-subtitle-format	org-latex-subtitle-format
:latex-subtitle-separate	org-latex-subtitle-separate
:latex-table-scientific-notation	org-latex-table-scientific-notation
:latex-tables-booktabs	org-latex-tables-booktabs
:latex-tables-centered	org-latex-tables-centered
:latex-text-markup-alist	org-latex-text-markup-alist
:latex-title-command	org-latex-title-command
:latex-toc-command	org-latex-toc-command

Markdown specific properties

:md-footnote-format	org-md-footnote-format
:md-footnotes-section	org-md-footnotes-section
:md-headline-style	org-md-headline-style

ODT specific properties

:odt-content-template-file	org-odt-content-template-file
:odt-display-outline-level	org-odt-display-outline-level
:odt-fontify-srcblocks	org-odt-fontify-srcblocks
:odt-format-drawer-function	org-odt-format-drawer-function
:odt-format-headline-function	org-odt-format-headline-function
:odt-format-inlinetask-function	org-odt-format-inlinetask-function
:odt-inline-formula-rules	org-odt-inline-formula-rules
:odt-inline-image-rules	org-odt-inline-image-rules
:odt-pixels-per-inch	org-odt-pixels-per-inch
:odt-styles-file	org-odt-styles-file
:odt-table-styles	org-odt-table-styles
:odt-use-date-fields	org-odt-use-date-fields

Texinfo specific properties

`:texinfo-active-timestamp-format`	`org-texinfo-active-timestamp-format`
`:texinfo-classes`	`org-texinfo-classes`
`:texinfo-class`	`org-texinfo-default-class`
`:texinfo-table-default-markup`	`org-texinfo-table-default-markup`
`:texinfo-diary-timestamp-format`	`org-texinfo-diary-timestamp-format`
`:texinfo-filename`	`org-texinfo-filename`
`:texinfo-format-drawer-function`	`org-texinfo-format-drawer-function`
`:texinfo-format-headline-function`	`org-texinfo-format-headline-function`
`:texinfo-format-inlinetask-function`	`org-texinfo-format-inlinetask-function`
`:texinfo-inactive-timestamp-format`	`org-texinfo-inactive-timestamp-format`
`:texinfo-link-with-unknown-path-format`	`org-texinfo-link-with-unknown-path-forma`
`:texinfo-node-description-column`	`org-texinfo-node-description-column`
`:texinfo-table-scientific-notation`	`org-texinfo-table-scientific-notation`
`:texinfo-tables-verbatim`	`org-texinfo-tables-verbatim`
`:texinfo-text-markup-alist`	`org-texinfo-text-markup-alist`

13.1.6 Links between published files

To create a link from one Org file to another, you would use something like '`[[file:foo.org][The foo]]`' or simply '`file:foo.org`' (see Section 4.3 [External links], page 39). When published, this link becomes a link to `foo.html`. You can thus interlink the pages of your "org web" project and the links will work as expected when you publish them to HTML. If you also publish the Org source file and want to link to it, use an `http:` link instead of a `file:` link, because `file:` links are converted to link to the corresponding `html` file.

You may also link to related files, such as images. Provided you are careful with relative file names, and provided you have also configured Org to upload the related files, these links will work too. See Section 13.3.2 [Complex example], page 196, for an example of this usage.

Eventually, links between published documents can contain some search options (see Section 4.7 [Search options], page 45), which will be resolved to the appropriate location in the linked file. For example, once published to HTML, the following links all point to a dedicated anchor in `foo.html`.

```
[[file:foo.org::*heading]]
[[file:foo.org::#custom-id]]
[[file:foo.org::target]]
```

13.1.7 Generating a sitemap

The following properties may be used to control publishing of a map of files for a given project.

`:auto-sitemap`
: When non-`nil`, publish a sitemap during `org-publish-current-project` or `org-publish-all`.

`:sitemap-filename`
: Filename for output of sitemap. Defaults to `sitemap.org` (which becomes `sitemap.html`).

`:sitemap-title`
: Title of sitemap page. Defaults to name of file.

`:sitemap-format-entry`
: With this option one can tell how a site-map entry is formatted in the site-map. It is a function called with three arguments: the file or directory name relative to base directory of the project, the site-map style and the current project. It is expected to return a string. Default value turns file names into links and use document titles as descriptions. For specific formatting needs, one can use `org-publish-find-date`, `org-publish-find-title` and `org-publish-find-property`, to retrieve additional information about published documents.

`:sitemap-function`
: Plug-in function to use for generation of the sitemap. It is called with two arguments: the title of the site-map and a representation of the files and directories involved in the project as a radio list (see Section A.6.4 [Radio lists], page 246). The latter can further be transformed using `org-list-to-generic`, `org-list-to-subtree` and alike. Default value generates a plain list of links to all files in the project.

`:sitemap-sort-folders`
: Where folders should appear in the sitemap. Set this to `first` (default) or `last` to display folders first or last, respectively. When set to `ignore`, folders are ignored altogether. Any other value will mix files and folders. This variable has no effect when site-map style is `tree`.

`:sitemap-sort-files`
: How the files are sorted in the site map. Set this to `alphabetically` (default), `chronologically` or `anti-chronologically`. `chronologically` sorts the files with older date first while `anti-chronologically` sorts the files with newer date first. `alphabetically` sorts the files alphabetically. The date of a file is retrieved with `org-publish-find-date`.

`:sitemap-ignore-case`
: Should sorting be case-sensitive? Default `nil`.

`:sitemap-date-format`
: Format string for the `format-time-string` function that tells how a sitemap entry's date is to be formatted. This property bypasses `org-publish-sitemap-date-format` which defaults to `%Y-%m-%d`.

13.1.8 Generating an index

Org mode can generate an index across the files of a publishing project.

`:makeindex` When non-`nil`, generate in index in the file `theindex.org` and publish it as `theindex.html`.

The file will be created when first publishing a project with the `:makeindex` set. The file only contains a statement `#+INCLUDE: "theindex.inc"`. You can then build around this include statement by adding a title, style information, etc.

Index entries are specified with `#+INDEX` keyword. An entry that contains an exclamation mark will create a sub item.

```
* Curriculum Vitae
#+INDEX: CV
#+INDEX: Application!CV
```

13.2 Uploading files

For those people already utilizing third party sync tools such as `rsync` or `unison`, it might be preferable not to use the built in *remote* publishing facilities of Org mode which rely heavily on Tramp. Tramp, while very useful and powerful, tends not to be so efficient for multiple file transfer and has been known to cause problems under heavy usage.

Specialized synchronization utilities offer several advantages. In addition to timestamp comparison, they also do content and permissions/attribute checks. For this reason you might prefer to publish your web to a local directory (possibly even *in place* with your Org files) and then use `unison` or `rsync` to do the synchronization with the remote host.

Since Unison (for example) can be configured as to which files to transfer to a certain remote destination, it can greatly simplify the project publishing definition. Simply keep all files in the correct location, process your Org files with `org-publish` and let the synchronization tool do the rest. You do not need, in this scenario, to include attachments such as `jpg`, `css` or `gif` files in the project definition since the 3rd party tool syncs them.

Publishing to a local directory is also much faster than to a remote one, so that you can afford more easily to republish entire projects. If you set `org-publish-use-timestamps-flag` to `nil`, you gain the main benefit of re-including any changed external files such as source example files you might include with `#+INCLUDE:`. The timestamp mechanism in Org is not smart enough to detect if included files have been modified.

13.3 Sample configuration

Below we provide two example configurations. The first one is a simple project publishing only a set of Org files. The second example is more complex, with a multi-component project.

13.3.1 Example: simple publishing configuration

This example publishes a set of Org files to the `public_html` directory on the local machine.

```
(setq org-publish-project-alist
      '(("org"
         :base-directory "~/org/"
         :publishing-directory "~/public_html"
         :publishing-function org-html-publish-to-html
         :section-numbers nil
```

```
:with-toc nil
:html-head "<link rel=\"stylesheet\"
           href=\"../other/mystyle.css\"
           type=\"text/css\"/>")))
```

13.3.2 Example: complex publishing configuration

This more complicated example publishes an entire website, including Org files converted to HTML, image files, Emacs Lisp source code, and style sheets. The publishing directory is remote and private files are excluded.

To ensure that links are preserved, care should be taken to replicate your directory structure on the web server, and to use relative file paths. For example, if your Org files are kept in `~/org` and your publishable images in `~/images`, you would link to an image with

```
file:../images/myimage.png
```

On the web server, the relative path to the image should be the same. You can accomplish this by setting up an "images" folder in the right place on the web server, and publishing images to it.

```
(setq org-publish-project-alist
      '(("orgfiles"
         :base-directory "~/org/"
         :base-extension "org"
         :publishing-directory "/ssh:user@host:~/html/notebook/"
         :publishing-function org-html-publish-to-html
         :exclude "PrivatePage.org"   ;; regexp
         :headline-levels 3
         :section-numbers nil
         :with-toc nil
         :html-head "<link rel=\"stylesheet\"
                 href=\"../other/mystyle.css\" type=\"text/css\"/>"
         :html-preamble t)

        ("images"
         :base-directory "~/images/"
         :base-extension "jpg\\|gif\\|png"
         :publishing-directory "/ssh:user@host:~/html/images/"
         :publishing-function org-publish-attachment)

        ("other"
         :base-directory "~/other/"
         :base-extension "css\\|el"
         :publishing-directory "/ssh:user@host:~/html/other/"
         :publishing-function org-publish-attachment)
        ("website" :components ("orgfiles" "images" "other"))))
```

13.4 Triggering publication

Once properly configured, Org can publish with the following commands:

C-c C-e P x `org-publish`
: Prompt for a specific project and publish all files that belong to it.

C-c C-e P p `org-publish-current-project`
: Publish the project containing the current file.

C-c C-e P f `org-publish-current-file`
: Publish only the current file.

C-c C-e P a `org-publish-all`
: Publish every project.

Org uses timestamps to track when a file has changed. The above functions normally only publish changed files. You can override this and force publishing of all files by giving a prefix argument to any of the commands above, or by customizing the variable `org-publish-use-timestamps-flag`. This may be necessary in particular if files include other files via `#+SETUPFILE:` or `#+INCLUDE:`.

14 Working with source code

Source code here refers to any code typed in Org mode documents. Org can manage source code in any Org file once such code is tagged with begin and end markers. Working with source code begins with tagging source code blocks. Tagged 'src' code blocks are not restricted to the preamble or the end of an Org document; they can go anywhere—with a few exceptions, such as not inside comments and fixed width areas. Here's a sample 'src' code block in emacs-lisp:

```
#+BEGIN_SRC emacs-lisp
  (defun org-xor (a b)
     "Exclusive or."
     (if a (not b) b))
#+END_SRC
```

Org can take the code in the block between the '#+BEGIN_SRC' and '#+END_SRC' tags, and format, compile, execute, and show the results. Org can simplify many housekeeping tasks essential to modern code maintenance. That's why these blocks in Org mode literature are sometimes referred to as 'live code' blocks (as compared to the static text and documentation around it). Users can control how 'live' they want each block by tweaking the headers for compiling, execution, extraction.

Org's 'src' code block type is one of many block types, such as quote, export, verse, latex, example, and verbatim. This section pertains to 'src' code blocks between '#+BEGIN_SRC' and '#+END_SRC'

For editing 'src' code blocks, Org provides native Emacs major-modes. That leverages the latest Emacs features for that source code language mode.

For exporting, Org can then extract 'src' code blocks into compilable source files (in a conversion process known as *tangling* in literate programming terminology).

For publishing, Org's back-ends can handle the 'src' code blocks and the text for output to a variety of formats with native syntax highlighting.

For executing the source code in the 'src' code blocks, Org provides facilities that glue the tasks of compiling, collecting the results of the execution, and inserting them back to the Org file. Besides text output, results may include links to other data types that Emacs can handle: audio, video, and graphics.

An important feature of Org's execution of the 'src' code blocks is passing variables, functions, and results between 'src' blocks. Such interoperability uses a common syntax even if these 'src' blocks are in different source code languages. The integration extends to linking the debugger's error messages to the line in the 'src' code block in the Org file. That should partly explain why this functionality by the original contributors, Eric Schulte and Dan Davison, was called 'Org Babel'.

In literate programming, the main appeal is code and documentation co-existing in one file. Org mode takes this several steps further. First by enabling execution, and then by inserting results of that execution back into the Org file. Along the way, Org provides extensive formatting features, including handling tables. Org handles multiple source code languages in one file, and provides a common syntax for passing variables, functions, and results between 'src' code blocks.

Org mode fulfills the promise of easy verification and maintenance of publishing reproducible research by keeping all these in the same file: text, data, code, configuration settings of the execution environment, the results of the execution, and associated narratives, claims, references, and internal and external links.

Details of Org's facilities for working with source code are shown next.

14.1 Structure of code blocks

Org offers two ways to structure source code in Org documents: in a 'src' block, and directly inline. Both specifications are shown below.

A 'src' block conforms to this structure:

```
#+NAME: <name>
#+BEGIN_SRC <language> <switches> <header arguments>
  <body>
#+END_SRC
```

Org mode's templates system (see Section 15.2 [Easy templates], page 228) speeds up creating 'src' code blocks with just three keystrokes. Do not be put-off by having to remember the source block syntax. Org also works with other completion systems in Emacs, some of which predate Org and have custom domain-specific languages for defining templates. Regular use of templates reduces errors, increases accuracy, and maintains consistency.

An inline code block conforms to this structure:

```
src_<language>{<body>}
```

or

```
src_<language>[<header arguments>]{<body>}
```

#+NAME: <name>

Optional. Names the 'src' block so it can be called, like a function, from other 'src' blocks or inline blocks to evaluate or to capture the results. Code from other blocks, other files, and from table formulas (see Section 3.5 [The spreadsheet], page 24) can use the name to reference a 'src' block. This naming serves the same purpose as naming Org tables. Org mode requires unique names. For duplicate names, Org mode's behavior is undefined.

#+BEGIN_SRC
#+END_SRC

Mandatory. They mark the start and end of a block that Org requires. The #+BEGIN_SRC line takes additional arguments, as described next.

<language>

Mandatory for live code blocks. It is the identifier of the source code language in the block. See Section 14.7 [Languages], page 204, for identifiers of supported languages.

<switches>

Optional. Switches provide finer control of the code execution, export, and format (see the discussion of switches in Section 11.5 [Literal examples], page 131)

`<header arguments>`
Optional. Heading arguments control many aspects of evaluation, export and tangling of code blocks (see Section 14.8 [Header arguments], page 205). Using Org's properties feature, header arguments can be selectively applied to the entire buffer or specific sub-trees of the Org document.

`source code, header arguments`
`<body>` Source code in the dialect of the specified language identifier.

14.2 Editing source code

C-c ' for editing the current code block. It opens a new major-mode edit buffer containing the body of the 'src' code block, ready for any edits. *C-c '* again to close the buffer and return to the Org buffer.

C-x C-s saves the buffer and updates the contents of the Org buffer.

Set `org-edit-src-auto-save-idle-delay` to save the base buffer after a certain idle delay time.

Set `org-edit-src-turn-on-auto-save` to auto-save this buffer into a separate file using `auto-save-mode`.

C-c ' to close the major-mode buffer and return back to the Org buffer.

While editing the source code in the major-mode, the `org-src-mode` minor mode remains active. It provides these customization variables as described below. For even more variables, look in the customization group `org-edit-structure`.

`org-src-lang-modes`
If an Emacs major-mode named `<lang>-mode` exists, where `<lang>` is the language identifier from code block's header line, then the edit buffer uses that major-mode. Use this variable to arbitrarily map language identifiers to major modes.

`org-src-window-setup`
For specifying Emacs window arrangement when the new edit buffer is created.

`org-src-preserve-indentation`
Default is `nil`. Source code is indented. This indentation applies during export or tangling, and depending on the context, may alter leading spaces and tabs. When non-`nil`, source code is aligned with the leftmost column. No lines are modified during export or tangling, which is very useful for white-space sensitive languages, such as Python.

`org-src-ask-before-returning-to-edit-buffer`
When `nil`, Org returns to the edit buffer without further prompts. The default prompts for a confirmation.

Set `org-src-fontify-natively` to non-`nil` to turn on native code fontification in the *Org* buffer. Fontification of 'src' code blocks can give visual separation of text and code on the display page. To further customize the appearance of `org-block` for specific languages, customize `org-src-block-faces`. The following example shades the background of regular blocks, and colors source blocks only for Python and Emacs-Lisp languages.

```
(require 'color)
(set-face-attribute 'org-block nil :background
                    (color-darken-name
                     (face-attribute 'default :background) 3))

(setq org-src-block-faces '(("emacs-lisp" (:background "#EEE2FF"))
                            ("python" (:background "#E5FFB8"))))
```

14.3 Exporting code blocks

Org can flexibly export just the *code* from the code blocks, just the *results* of evaluation of the code block, *both* the code and the results of the code block evaluation, or *none*. Org defaults to exporting *code* for most languages. For some languages, such as `ditaa`, Org defaults to *results*. To export just the body of code blocks, see Section 11.5 [Literal examples], page 131. To selectively export sub-trees of an Org document, see Chapter 12 [Exporting], page 137.

The `:exports` header arguments control exporting code blocks only and not inline code:

Header arguments:

`:exports code`
: This is the default for most languages where the body of the code block is exported. See Section 11.5 [Literal examples], page 131 for more.

`:exports results`
: On export, Org includes only the results and not the code block. After each evaluation, Org inserts the results after the end of code block in the Org buffer. By default, Org replaces any previous results. Org can also append results.

`:exports both`
: Org exports both the code block and the results.

`:exports none`
: Org does not export the code block nor the results.

To stop Org from evaluating code blocks to speed exports, use the header argument `:eval never-export` (see Section 14.8.2.25 [eval], page 222). To stop Org from evaluating code blocks for greater security, set the `org-export-use-babel` variable to `nil`, but understand that header arguments will have no effect.

Turning off evaluation comes in handy when batch processing. For example, markup languages for wikis, which have a high risk of untrusted code. Stopping code block evaluation also stops evaluation of all header arguments of the code block. This may not be desirable in some circumstances. So during export, to allow evaluation of just the header arguments but not any code evaluation in the source block, set `:eval never-export` (see Section 14.8.2.25 [eval], page 222).

Org never evaluates code blocks in commented sub-trees when exporting (see Section 12.6 [Comment lines], page 144). On the other hand, Org does evaluate code blocks in sub-trees excluded from export (see Section 12.2 [Export settings], page 138).

14.4 Extracting source code

Extracting source code from code blocks is a basic task in literate programming. Org has features to make this easy. In literate programming parlance, documents on creation are *woven* with code and documentation, and on export, the code is *tangled* for execution by a computer. Org facilitates weaving and tangling for producing, maintaining, sharing, and exporting literate programming documents. Org provides extensive customization options for extracting source code.

When Org tangles 'src' code blocks, it expands, merges, and transforms them. Then Org recomposes them into one or more separate files, as configured through the options. During this *tangling* process, Org expands variables in the source code, and resolves any Noweb style references (see Section 14.10 [Noweb reference syntax], page 225).

Header arguments

`:tangle no`
: By default, Org does not tangle the 'src' code block on export.

`:tangle yes`
: Org extracts the contents of the code block for the tangled output. By default, the output file name is the same as the Org file but with a file extension derived from the language identifier of the 'src' code block.

`:tangle filename`
: Override the default file name with this one for the tangled output.

Functions

`org-babel-tangle`
: Tangle the current file. Bound to *C-c C-v t*.

  With prefix argument only tangle the current 'src' code block.

`org-babel-tangle-file`
: Choose a file to tangle. Bound to *C-c C-v f*.

Hooks

`org-babel-post-tangle-hook`
: This hook runs from within code tangled by `org-babel-tangle`, making it suitable for post-processing, compilation, and evaluation of code in the tangled files.

Jumping between code and Org

Debuggers normally link errors and messages back to the source code. But for tangled files, we want to link back to the Org file, not to the tangled source file. To make this extra jump, Org uses `org-babel-tangle-jump-to-org` function with two additional source code block header arguments: One, set `padline` (see Section 14.8.2.12 [padline], page 215) to true (the default setting). Two, set `comments` (see Section 14.8.2.11 [comments], page 215) to `link`, which makes Org insert links to the Org file.

14.5 Evaluating code blocks

A note about security: With code evaluation comes the risk of harm. Org safeguards by prompting for user's permission before executing any code in the source block. To customize this safeguard (or disable it) see Section 15.4 [Code evaluation security], page 229.

Org captures the results of the 'src' code block evaluation and inserts them in the Org file, right after the 'src' code block. The insertion point is after a newline and the `#+RESULTS` label. Org creates the `#+RESULTS` label if one is not already there.

By default, Org enables only `emacs-lisp` 'src' code blocks for execution. See Section 14.7 [Languages], page 204 for identifiers to enable other languages.

Org provides many ways to execute 'src' code blocks. *C-c C-c* or *C-c C-v e* with the point on a 'src' code block[1] calls the `org-babel-execute-src-block` function, which executes the code in the block, collects the results, and inserts them in the buffer.

By calling a named code block[2] from an Org mode buffer or a table. Org can call the named 'src' code blocks from the current Org mode buffer or from the "Library of Babel" (see Section 14.6 [Library of Babel], page 204). Whether inline syntax or the `#+CALL:` syntax is used, the result is wrapped based on the variable `org-babel-inline-result-wrap`, which by default is set to `"=%s="` to produce verbatim text suitable for markup.

The syntax for `#+CALL:` is

```
#+CALL: <name>(<arguments>)
#+CALL: <name>[<inside header arguments>](<arguments>) <end header arguments
```

The syntax for inline named code block is

```
... call_<name>(<arguments>) ...
... call_<name>[<inside header arguments>](<arguments>)[<end header argument
```

`<name>` This is the name of the code block to be evaluated (see Section 14.1 [Structure of code blocks], page 199).

`<arguments>`
Org passes arguments to the code block using standard function call syntax. For example, a `#+CALL:` line that passes '4' to a code block named `double`, which declares the header argument `:var n=2`, would be written as `#+CALL: double(n=4)`. Note how this function call syntax is different from the header argument syntax.

`<inside header arguments>`
Org passes inside header arguments to the named 'src' code block using the header argument syntax. Inside header arguments apply to code block evaluation. For example, `[:results output]` collects results printed to `STDOUT` during code execution of that block. Note how this header argument syntax is different from the function call syntax.

---

[1] The option `org-babel-no-eval-on-ctrl-c-ctrl-c` can be used to remove code evaluation from the *C-c C-c* key binding.

[2] Actually, the constructs call_<name>() and src_<lang>{} are not evaluated when they appear in a keyword line (i.e. lines starting with `#+KEYWORD:`, see Section 15.6 [In-buffer settings], page 230).

`<end header arguments>`

End header arguments affect the results returned by the code block. For example, `:results html` wraps the results in a `BEGIN_EXPORT html` block before inserting the results in the Org buffer.

For more examples of header arguments for `#+CALL:` lines, see [Arguments in function calls], page 207.

14.6 Library of Babel

The "Library of Babel" is a collection of code blocks. Like a function library, these code blocks can be called from other Org files. A collection of useful code blocks is available on Worg. For remote code block evaluation syntax, see Section 14.5 [Evaluating code blocks], page 203.

For any user to add code to the library, first save the code in regular '`src`' code blocks of an Org file, and then load the Org file with `org-babel-lob-ingest`, which is bound to *C-c C-v i*.

14.7 Languages

Org supports the following languages for the '`src`' code blocks:

Language	**Identifier**	**Language**	**Identifier**
Asymptote	asymptote	Awk	awk
C	C	C++	C++
Clojure	clojure	CSS	css
D	d	ditaa	ditaa
Graphviz	dot	Emacs Calc	calc
Emacs Lisp	emacs-lisp	Fortran	fortran
gnuplot	gnuplot	Haskell	haskell
Java	java	Javascript	js
LaTeX	latex	Ledger	ledger
Lisp	lisp	Lilypond	lilypond
Lua	lua	MATLAB	matlab
Mscgen	mscgen	Objective Caml	ocaml
Octave	octave	Org mode	org
Oz	oz	Perl	perl
Plantuml	plantuml	Processing.js	processing
Python	python	R	R
Ruby	ruby	Sass	sass
Scheme	scheme	GNU Screen	screen
Sed	sed	shell	sh
SQL	sql	SQLite	sqlite
Vala	vala		

Additional documentation for some languages are at `http://orgmode.org/worg/org-contrib/babel/languages.html`.

By default, only `emacs-lisp` is enabled for evaluation. To enable or disable other languages, customize the `org-babel-load-languages` variable either through the Emacs customization interface, or by adding code to the init file as shown next:

In this example, evaluation is disabled for `emacs-lisp`, and enabled for `R`.

```
(org-babel-do-load-languages
 'org-babel-load-languages
 '((emacs-lisp . nil)
   (R . t)))
```

Note that this is not the only way to enable a language. Org also enables languages when loaded with `require` statement. For example, the following enables execution of `clojure` code blocks:

```
(require 'ob-clojure)
```

14.8 Header arguments

Details of configuring header arguments are shown here.

14.8.1 Using header arguments

Since header arguments can be set in several ways, Org prioritizes them in case of overlaps or conflicts by giving local settings a higher priority. Header values in function calls, for example, override header values from global defaults.

System-wide header arguments

System-wide values of header arguments can be specified by adapting the `org-babel-default-header-args` variable:

```
:session    => "none"
:results    => "replace"
:exports    => "code"
:cache      => "no"
:noweb      => "no"
```

This example sets `:noweb` header arguments to `yes`, which makes Org expand `:noweb` references by default.

```
(setq org-babel-default-header-args
      (cons '(:noweb . "yes")
            (assq-delete-all :noweb org-babel-default-header-args)))
```

Language-specific header arguments

Each language can have separate default header arguments by customizing the variable `org-babel-default-header-args:<lang>`, where `<lang>` is the name of the language. For details, see the language-specific online documentation at http://orgmode.org/worg/org-contrib/babel.

Header arguments in Org mode properties

For header arguments applicable to the buffer, use `#+PROPERTY:` lines anywhere in the Org mode file (see Section 7.1 [Property syntax], page 64).

The following example sets only for 'R' code blocks to **session**, making all the 'R' code blocks execute in the same session. Setting **results** to **silent** ignores the results of executions for all blocks, not just 'R' code blocks; no results inserted for any block.

```
#+PROPERTY: header-args:R  :session *R*
#+PROPERTY: header-args    :results silent
```

Header arguments set through Org's property drawers (see Section 7.1 [Property syntax], page 64) apply at the sub-tree level on down. Since these property drawers can appear anywhere in the file hierarchy, Org uses outermost call or source block to resolve the values. Org ignores `org-use-property-inheritance` setting.

In this example, `:cache` defaults to `yes` for all code blocks in the sub-tree starting with '`sample header`'.

```
* sample header
  :PROPERTIES:
  :header-args:    :cache yes
  :END:
```

Properties defined through `org-set-property` function, bound to *C-c C-x p*, apply to all active languages. They override properties set in `org-babel-default-header-args`.

Language-specific mode properties

Language-specific header arguments are also read from properties `header-args:<lang>` where `<lang>` is the language identifier. For example,

```
* Heading
  :PROPERTIES:
  :header-args:clojure:    :session *clojure-1*
  :header-args:R:          :session *R*
  :END:
** Subheading
  :PROPERTIES:
  :header-args:clojure:    :session *clojure-2*
  :END:
```

would force separate sessions for clojure blocks in Heading and Subheading, but use the same session for all 'R' blocks. Blocks in Subheading inherit settings from Heading.

Code block specific header arguments

Header arguments are most commonly set at the '`src`' code block level, on the `#+BEGIN_SRC` line. Arguments set at this level take precedence over those set in the `org-babel-default-header-args` variable, and also those set as header properties.

In the following example, setting `results` to `silent` makes it ignore results of the code execution. Setting `:exports` to `code` exports only the body of the '`src`' code block to HTML or LaTeX.:

```
#+NAME: factorial
#+BEGIN_SRC haskell :results silent :exports code :var n=0
fac 0 = 1
fac n = n * fac (n-1)
#+END_SRC
```

The same header arguments in an inline 'src' code block:

```
src_haskell[:exports both]{fac 5}
```

Code block header arguments can span multiple lines using #+HEADER: on each line. Note that Org currently accepts the plural spelling of #+HEADER: only as a convenience for backward-compatibility. It may be removed at some point.

Multi-line header arguments on an unnamed 'src' code block:

```
#+HEADER: :var data1=1
#+BEGIN_SRC emacs-lisp :var data2=2
   (message "data1:%S, data2:%S" data1 data2)
#+END_SRC

#+RESULTS:
: data1:1, data2:2
```

Multi-line header arguments on a named 'src' code block:

```
#+NAME: named-block
#+HEADER: :var data=2
#+BEGIN_SRC emacs-lisp
  (message "data:%S" data)
#+END_SRC

#+RESULTS: named-block
  : data:2
```

Arguments in function calls

Header arguments in function calls are the most specific and override all other settings in case of an overlap. They get the highest priority. Two #+CALL: examples are shown below. For the complete syntax of #+CALL: lines, see Section 14.5 [Evaluating code blocks], page 203.

In this example, :exports results header argument is applied to the evaluation of the #+CALL: line.

```
#+CALL: factorial(n=5) :exports results
```

In this example, :session special header argument is applied to the evaluation of factorial code block.

```
#+CALL: factorial[:session special](n=5)
```

14.8.2 Specific header arguments

Org comes with many header arguments common to all languages. New header arguments are added for specific languages as they become available for use in 'src' code blocks. A header argument is specified with an initial colon followed by the argument's name in lowercase. Common header arguments are:

For language-specific header arguments, see Section 14.7 [Languages], page 204.

14.8.2.1 :var

Use `:var` for passing arguments to 'src' code blocks. The specifics of variables in 'src' code blocks vary by the source language and are covered in the language-specific documentation. The syntax for `:var`, however, is the same for all languages. This includes declaring a variable, and assigning a default value.

Arguments can take values as literals, or as references, or even as Emacs Lisp code (see Section 14.8.2.1 [var], page 208). References are names from the Org file from the lines `#+NAME:` or `#+RESULTS:`. References can also refer to tables, lists, `#+BEGIN_EXAMPLE` blocks, other types of 'src' code blocks, or the results of execution of 'src' code blocks.

For better performance, Org can cache results of evaluations. But caching comes with severe limitations (see Section 14.8.2.18 [cache], page 218).

Argument values are indexed like arrays (see Section 14.8.2.1 [var], page 208).

The following syntax is used to pass arguments to 'src' code blocks using the `:var` header argument.

```
:var name=assign
```

The `assign` is a literal value, such as a string '"string"', a number '9', a reference to a table, a list, a literal example, another code block (with or without arguments), or the results from evaluating a code block.

Here are examples of passing values by reference:

table an Org mode table named with either a `#+NAME:` line

```
#+NAME: example-table
| 1 |
| 2 |
| 3 |
| 4 |

#+NAME: table-length
#+BEGIN_SRC emacs-lisp :var table=example-table
(length table)
#+END_SRC

#+RESULTS: table-length
: 4
```

list a simple list named with a `#+NAME:` line. Note that only the top level list items are passed along. Nested list items are ignored.

```
#+NAME: example-list
  - simple
    - not
    - nested
  - list

#+BEGIN_SRC emacs-lisp :var x=example-list
  (print x)
#+END_SRC
```

```
#+RESULTS:
| simple | list |
```

code block without arguments

a code block name (from the example above), as assigned by `#+NAME:`, optionally followed by parentheses

```
#+BEGIN_SRC emacs-lisp :var length=table-length()
(* 2 length)
#+END_SRC

#+RESULTS:
: 8
```

code block with arguments

a '`src`' code block name, as assigned by `#+NAME:`, followed by parentheses and optional arguments passed within the parentheses following the '`src`' code block name using standard function call syntax

```
#+NAME: double
#+BEGIN_SRC emacs-lisp :var input=8
(* 2 input)
#+END_SRC

#+RESULTS: double
: 16

#+NAME: squared
#+BEGIN_SRC emacs-lisp :var input=double(input=2)
(* input input)
#+END_SRC

#+RESULTS: squared
: 4
```

literal example

a literal example block named with a `#+NAME:` line

```
#+NAME: literal-example
#+BEGIN_EXAMPLE
A literal example
on two lines
#+END_EXAMPLE

#+NAME: read-literal-example
#+BEGIN_SRC emacs-lisp :var x=literal-example
  (concatenate 'string x " for you.")
#+END_SRC

#+RESULTS: read-literal-example
```

```
: A literal example
: on two lines for you.
```

Indexable variable values

Indexing variable values enables referencing portions of a variable. Indexes are 0 based with negative values counting backwards from the end. If an index is separated by `,`s then each subsequent section will index as the next dimension. Note that this indexing occurs *before* other table-related header arguments are applied, such as `:hlines`, `:colnames` and `:rownames`. The following example assigns the last cell of the first row the table `example-table` to the variable `data`:

```
#+NAME: example-table
| 1 | a |
| 2 | b |
| 3 | c |
| 4 | d |

#+BEGIN_SRC emacs-lisp :var data=example-table[0,-1]
  data
#+END_SRC

#+RESULTS:
: a
```

Ranges of variable values can be referenced using two integers separated by a `:`, in which case the entire inclusive range is referenced. For example the following assigns the middle three rows of `example-table` to `data`.

```
#+NAME: example-table
| 1 | a |
| 2 | b |
| 3 | c |
| 4 | d |
| 5 | 3 |

#+BEGIN_SRC emacs-lisp :var data=example-table[1:3]
  data
#+END_SRC

#+RESULTS:
| 2 | b |
| 3 | c |
| 4 | d |
```

To pick the entire range, use an empty index, or the single character `*`. `0:-1` does the same thing. Example below shows how to reference the first column only.

```
#+NAME: example-table
| 1 | a |
| 2 | b |
```

```
| 3 | c |
| 4 | d |

#+BEGIN_SRC emacs-lisp :var data=example-table[,0]
  data
#+END_SRC

#+RESULTS:
| 1 | 2 | 3 | 4 |
```

Index referencing can be used for tables and code blocks. Index referencing can handle any number of dimensions. Commas delimit multiple dimensions, as shown below.

```
#+NAME: 3D
#+BEGIN_SRC emacs-lisp
  '(((1  2  3)  (4  5  6)  (7  8  9))
    ((10 11 12) (13 14 15) (16 17 18))
    ((19 20 21) (22 23 24) (25 26 27)))
#+END_SRC

#+BEGIN_SRC emacs-lisp :var data=3D[1,,1]
  data
#+END_SRC

#+RESULTS:
| 11 | 14 | 17 |
```

Emacs Lisp evaluation of variables

Emacs lisp code can set the values for variables. To differentiate a value from lisp code, Org interprets any value starting with `(`, `[`, `'` or `` ` `` as Emacs Lisp code. The result of evaluating that code is then assigned to the value of that variable. The following example shows how to reliably query and pass file name of the Org mode buffer to a code block using headers. We need reliability here because the file's name could change once the code in the block starts executing.

```
#+BEGIN_SRC sh :var filename=(buffer-file-name) :exports both
  wc -w $filename
#+END_SRC
```

Note that values read from tables and lists will not be mistakenly evaluated as Emacs Lisp code, as illustrated in the following example.

```
#+NAME: table
| (a b c) |

#+HEADER: :var data=table[0,0]
#+BEGIN_SRC perl
  $data
#+END_SRC

#+RESULTS:
```

```
: (a b c)
```

14.8.2.2 `:results`

There are four classes of `:results` header arguments. Each 'src' code block can take only one option per class.

- **collection** for how the results should be collected from the 'src' code block
- **type** for which type of result the code block will return; affects how Org processes and inserts results in the Org buffer
- **format** for the result; affects how Org processes and inserts results in the Org buffer
- **handling** for processing results after evaluation of the 'src' code block

Collection

Collection options specify the results. Choose one of the options; they are mutually exclusive.

- `value` Default. Functional mode. Result is the value returned by the last statement in the 'src' code block. Languages like Python may require an explicit `return` statement in the 'src' code block. Usage example: `:results value`.
- `output` Scripting mode. Result is collected from STDOUT during execution of the code in the 'src' code block. Usage example: `:results output`.

Type

Type tells what result types to expect from the execution of the code block. Choose one of the options; they are mutually exclusive. The default behavior is to automatically determine the result type.

- `table`, `vector` Interpret the results as an Org table. If the result is a single value, create a table with one row and one column. Usage example: `:results value table`.
- `list` Interpret the results as an Org list. If the result is a single value, create a list of one element.
- `scalar`, `verbatim` Interpret literally and insert as quoted text. Do not create a table. Usage example: `:results value verbatim`.
- `file` Interpret as path to a file. Inserts a link to the file. Usage example: `:results value file`.

Format

Format pertains to the type of the result returned by the 'src' code block. Choose one of the options; they are mutually exclusive. The default follows from the type specified above.

- `raw` Interpreted as raw Org mode. Inserted directly into the buffer. Aligned if it is a table. Usage example: `:results value raw`.
- `org` Results enclosed in a `BEGIN_SRC org` block. For comma-escape, either *`TAB`* in the block, or export the file. Usage example: `:results value org`.
- `html` Results enclosed in a `BEGIN_EXPORT html` block. Usage example: `:results value html`.
- `latex` Results enclosed in a `BEGIN_EXPORT latex` block. Usage example: `:results value latex`.

- `code` Result enclosed in a '`src`' code block. Useful for parsing. Usage example: `:results value code`.
- `pp` Result converted to pretty-print source code. Enclosed in a '`src`' code block. Languages supported: Emacs Lisp, Python, and Ruby. Usage example: `:results value pp`.
- `drawer` Result wrapped in a RESULTS drawer. Useful for containing `raw` or `org` results for later scripting and automated processing. Usage example: `:results value drawer`.

Handling

Handling options after collecting the results.

- `silent` Do not insert results in the Org mode buffer, but echo them in the minibuffer. Usage example: `:results output silent`.
- `replace` Default. Insert results in the Org buffer. Remove previous results. Usage example: `:results output replace`.
- `append` Append results to the Org buffer. Latest results are at the bottom. Does not remove previous results. Usage example: `:results output append`.
- `prepend` Prepend results to the Org buffer. Latest results are at the top. Does not remove previous results. Usage example: `:results output prepend`.

14.8.2.3 :file

An external `:file` that saves the results of execution of the code block. The `:file` is either a file name or two strings, where the first is the file name and the second is the description. A link to the file is inserted. It uses an Org mode style `[[file:]]` link (see Section 4.1 [Link format], page 38). Some languages, such as '`R`', '`dot`', '`ditaa`', and '`gnuplot`', automatically wrap the source code in additional boilerplate code. Such code wrapping helps recreate the output, especially graphics output, by executing just the `:file` contents.

14.8.2.4 :file-desc

A description of the results file. Org uses this description for the link (see Section 4.1 [Link format], page 38) it inserts in the Org file. If the `:file-desc` has no value, Org will use file name for both the "link" and the "description" portion of the Org mode link.

14.8.2.5 :file-ext

File name extension for the output file. Org generates the file's complete name, and extension by combining `:file-ext`, `#+NAME:` of the source block, and the Section 14.8.2.6 [output-dir], page 213 header argument. To override this auto generated file name, use the `:file` header argument.

14.8.2.6 :output-dir

Specifies the `:output-dir` for the results file. Org accepts an absolute path (beginning with `/`) or a relative directory (without `/`). The value can be combined with `#+NAME:` of the source block and Section 14.8.2.3 [file], page 213 or Section 14.8.2.5 [file-ext], page 213 header arguments.

14.8.2.7 `:dir` and remote execution

While the `:file` header argument can be used to specify the path to the output file, `:dir` specifies the default directory during '`src`' code block execution. If it is absent, then the directory associated with the current buffer is used. In other words, supplying `:dir path` temporarily has the same effect as changing the current directory with *M-x cd path RET*, and then not supplying `:dir`. Under the surface, `:dir` simply sets the value of the Emacs variable `default-directory`.

When using `:dir`, relative paths (for example, `:file myfile.jpg` or `:file results/myfile.jpg`) become relative to the default directory.

For example, to save the plot file in the '`Work`' folder of the home directory (notice tilde is expanded):

```
#+BEGIN_SRC R :file myplot.png :dir ~/Work
matplot(matrix(rnorm(100), 10), type="l")
#+END_SRC
```

Remote execution

To evaluate the '`src`' code block on a remote machine, supply a remote s directory name using '`Tramp`' syntax. For example:

```
#+BEGIN_SRC R :file plot.png :dir /scp:dand@yakuba.princeton.edu:
plot(1:10, main=system("hostname", intern=TRUE))
#+END_SRC
```

Org first captures the text results as usual for insertion in the Org file. Then Org also inserts a link to the remote file, thanks to Emacs '`Tramp`'. Org constructs the remote path to the file name from `:dir` and `default-directory`, as illustrated here:

```
[[file:/scp:dand@yakuba.princeton.edu:/home/dand/plot.png][plot.png]]
```

Some more warnings

- When `:dir` is used with `:session`, Org sets the starting directory for a new session. But Org will not alter the directory of an already existing session.
- Do not use `:dir` with `:exports results` or with `:exports both` to avoid Org inserting incorrect links to remote files. That is because Org does not expand `default directory` to avoid some underlying portability issues.

14.8.2.8 `:exports`

The `:exports` header argument is to specify if that part of the Org file is exported to, say, HTML or LaTeX formats. Note that `:exports` affects only '`src`' code blocks and not inline code.

- `code` The default. The body of code is included into the exported file. Example: `:exports code`.
- `results` The results of evaluation of the code is included in the exported file. Example: `:exports results`.
- `both` Both the code and results of evaluation are included in the exported file. Example: `:exports both`.

- `none` Neither the code nor the results of evaluation is included in the exported file. Whether the code is evaluated at all depends on other options. Example: `:exports none`.

14.8.2.9 :tangle

The `:tangle` header argument specifies if the 'src' code block is exported to source file(s).

- `tangle` Export the 'src' code block to source file. The file name for the source file is derived from the name of the Org file, and the file extension is derived from the source code language identifier. Example: `:tangle yes`.
- `no` The default. Do not extract the code a source code file. Example: `:tangle no`.
- other Export the 'src' code block to source file whose file name is derived from any string passed to the `:tangle` header argument. Org derives the file name as being relative to the directory of the Org file's location. Example: `:tangle path`.

14.8.2.10 :mkdirp

The `:mkdirp` header argument creates parent directories for tangled files if the directory does not exist. `yes` enables directory creation and `no` inhibits directory creation.

14.8.2.11 :comments

Controls inserting comments into tangled files. These are above and beyond whatever comments may already exist in the 'src' code block.

- `no` The default. Do not insert any extra comments during tangling.
- `link` Wrap the 'src' code block in comments. Include links pointing back to the place in the Org file from where the code was tangled.
- `yes` Kept for backward compatibility; same as "link".
- `org` Nearest headline text from Org file is inserted as comment. The exact text that is inserted is picked from the leading context of the source block.
- `both` Includes both "link" and "org" comment options.
- `noweb` Includes "link" comment option, expands noweb references, and wraps them in link comments inside the body of the 'src' code block.

14.8.2.12 :padline

Control insertion of newlines to pad 'src' code blocks in the tangled file.

- `yes` Default. Insert a newline before and after each 'src' code block in the tangled file.
- `no` Do not insert newlines to pad the tangled 'src' code blocks.

14.8.2.13 :no-expand

By default Org expands 'src' code blocks during tangling. The `:no-expand` header argument turns off such expansions. Note that one side-effect of expansion by `org-babel-expand-src-block` also assigns values to `:var` (see Section 14.8.2.1 [var], page 208) variables. Expansions also replace Noweb references with their targets (see Section 14.10 [Noweb reference syntax], page 225). Some of these expansions may cause premature assignment, hence this option. This option makes a difference only for tangling. It has no effect when exporting since 'src' code blocks for execution have to be expanded anyway.

14.8.2.14 :session

The :session header argument is for running multiple source code blocks under one session. Org runs 'src' code blocks with the same session name in the same interpreter process.

- none Default. Each 'src' code block gets a new interpreter process to execute. The process terminates once the block is evaluated.
- other Any string besides none turns that string into the name of that session. For example, :session mysession names it 'mysession'. If :session has no argument, then the session name is derived from the source language identifier. Subsequent blocks with the same source code language use the same session. Depending on the language, state variables, code from other blocks, and the overall interpreted environment may be shared. Some interpreted languages support concurrent sessions when subsequent source code language blocks change session names.

14.8.2.15 :noweb

The :noweb header argument controls expansion of Noweb syntax references (see Section 14.10 [Noweb reference syntax], page 225). Expansions occur when source code blocks are evaluated, tangled, or exported.

- no Default. No expansion of Noweb syntax references in the body of the code when evaluating, tangling, or exporting.
- yes Expansion of Noweb syntax references in the body of the 'src' code block when evaluating, tangling, or exporting.
- tangle Expansion of Noweb syntax references in the body of the 'src' code block when tangling. No expansion when evaluating or exporting.
- no-export Expansion of Noweb syntax references in the body of the 'src' code block when evaluating or tangling. No expansion when exporting.
- strip-export Expansion of Noweb syntax references in the body of the 'src' code block when expanding prior to evaluating or tangling. Removes Noweb syntax references when exporting.
- eval Expansion of Noweb syntax references in the body of the 'src' code block only before evaluating.

Noweb prefix lines

Noweb insertions now honor prefix characters that appear before the Noweb syntax reference.

This behavior is illustrated in the following example. Because the <<example>> noweb reference appears behind the SQL comment syntax, each line of the expanded noweb reference will be commented.

With:

```
#+NAME: example
#+BEGIN_SRC text
this is the
multi-line body of example
#+END_SRC
```

this 'src' code block:

```
#+BEGIN_SRC sql :noweb yes
-- <<example>>
#+END_SRC
```

expands to:

```
-- this is the
-- multi-line body of example
```

Since this change will not affect noweb replacement text without newlines in them, inline noweb references are acceptable.

This feature can also be used for management of indentation in exported code snippets. With:

```
#+NAME: if-true
#+BEGIN_SRC python :exports none
print('Do things when True')
#+END_SRC

#+NAME: if-false
#+BEGIN_SRC python :exports none
print('Do things when False')
#+END_SRC
```

this 'src' code block:

```
#+BEGIN_SRC python :noweb yes :results output
if True:
    <<if-true>>
else:
    <<if-false>>
#+END_SRC
```

expands to:

```
if True:
    print('Do things when True')
else:
    print('Do things when False')
```

and evaluates to:

```
Do things when True
```

14.8.2.16 :noweb-ref

When expanding Noweb style references, Org concatenates '`src`' code blocks by matching the reference name to either the code block name or the `:noweb-ref` header argument.

For simple concatenation, set this `:noweb-ref` header argument at the sub-tree or file level. In the example Org file shown next, the body of the source code in each block is extracted for concatenation to a pure code file when tangled.

```
#+BEGIN_SRC sh :tangle yes :noweb yes :shebang #!/bin/sh
  <<fullest-disk>>
#+END_SRC
* the mount point of the fullest disk
```

```
  :PROPERTIES:
  :header-args: :noweb-ref fullest-disk
  :END:

** query all mounted disks
#+BEGIN_SRC sh
  df \
#+END_SRC

** strip the header row
#+BEGIN_SRC sh
  |sed '1d' \
#+END_SRC

** output mount point of fullest disk
#+BEGIN_SRC sh
  |awk '{if (u < +$5) {u = +$5; m = $6}} END {print m}'
#+END_SRC
```

14.8.2.17 :noweb-sep

By default a newline separates each noweb reference concatenation. To change this newline separator, edit the `:noweb-sep` (see Section 14.8.2.17 [noweb-sep], page 218) header argument.

14.8.2.18 :cache

The `:cache` header argument is for caching results of evaluating code blocks. Caching results can avoid re-evaluating '`src`' code blocks that have not changed since the previous run. To benefit from the cache and avoid redundant evaluations, the source block must have a result already present in the buffer, and neither the header arguments (including the value of `:var` references) nor the text of the block itself has changed since the result was last computed. This feature greatly helps avoid long-running calculations. For some edge cases, however, the cached results may not be reliable.

The caching feature is best for when '`src`' blocks are pure functions, that is functions that return the same value for the same input arguments (see Section 14.8.2.1 [var], page 208), and that do not have side effects, and do not rely on external variables other than the input arguments. Functions that depend on a timer, file system objects, and random number generators are clearly unsuitable for caching.

A note of warning: when `:cache` is used for a `:session`, caching may cause unexpected results.

When the caching mechanism tests for any source code changes, it will not expand Noweb style references (see Section 14.10 [Noweb reference syntax], page 225). For reasons why, see `http://thread.gmane.org/gmane.emacs.orgmode/79046`.

The `:cache` header argument can have one of two values: `yes` or `no`.

- `no` Default. No caching of results; '`src`' code block evaluated every time.
- `yes` Whether to run the code or return the cached results is determined by comparing the SHA1 hash value of the combined '`src`' code block and arguments passed to it.

This hash value is packed on the #+RESULTS: line from previous evaluation. When hash values match, Org does not evaluate the 'src' code block. When hash values mismatch, Org evaluates the 'src' code block, inserts the results, recalculates the hash value, and updates #+RESULTS: line.

In this example, both functions are cached. But caller runs only if the result from random has changed since the last run.

```
#+NAME: random
#+BEGIN_SRC R :cache yes
runif(1)
#+END_SRC

#+RESULTS[a2a72cd647ad44515fab62e144796432793d68e1]: random
0.4659510825295

#+NAME: caller
#+BEGIN_SRC emacs-lisp :var x=random :cache yes
x
#+END_SRC

#+RESULTS[bec9c8724e397d5df3b696502df3ed7892fc4f5f]: caller
0.254227238707244
```

14.8.2.19 :sep

The :sep header argument is the delimiter for saving results as tables to files (see Section 14.8.2.3 [file], page 213) external to Org mode. Org defaults to tab delimited output. The function, org-open-at-point, which is bound to *C-c C-o*, also uses :sep for opening tabular results.

14.8.2.20 :hlines

In-between each table row or below the table headings, sometimes results have horizontal lines, which are also known as hlines. The :hlines argument with the value yes accepts such lines. The default is no.

- no Strips horizontal lines from the input table. For most code, this is desirable, or else those hline symbols raise unbound variable errors.

  The default is :hlines no. The example shows hlines removed from the input table.

  ```
  #+NAME: many-cols
  | a | b | c |
  |---+---+---|
  | d | e | f |
  |---+---+---|
  | g | h | i |

  #+NAME: echo-table
  #+BEGIN_SRC python :var tab=many-cols
    return tab
  ```

```
#+END_SRC

#+RESULTS: echo-table
| a | b | c |
| d | e | f |
| g | h | i |
```

- `yes` For `:hlines yes`, the example shows hlines unchanged.

```
#+NAME: many-cols
| a | b | c |
|---+---+---|
| d | e | f |
|---+---+---|
| g | h | i |

#+NAME: echo-table
#+BEGIN_SRC python :var tab=many-cols :hlines yes
  return tab
#+END_SRC

#+RESULTS: echo-table
| a | b | c |
|---+---+---|
| d | e | f |
|---+---+---|
| g | h | i |
```

14.8.2.21 `:colnames`

The `:colnames` header argument accepts `yes`, `no`, or `nil` values. The default value is `nil`, which is unassigned. But this header argument behaves differently depending on the source code language.

- `nil` If an input table has column names (because the second row is an hline), then Org removes the column names, processes the table, puts back the column names, and then writes the table to the results block.

```
#+NAME: less-cols
| a |
|---|
| b |
| c |

#+NAME: echo-table-again
#+BEGIN_SRC python :var tab=less-cols
  return [[val + '*' for val in row] for row in tab]
#+END_SRC

#+RESULTS: echo-table-again
| a  |
```

```
|----|
| b* |
| c* |
```

Note that column names have to accounted for when using variable indexing (see Section 14.8.2.1 [var], page 208) because column names are not removed for indexing.

- `no` Do not pre-process column names.
- `yes` For an input table that has no hlines, process it like the `nil` value. That is, Org removes the column names, processes the table, puts back the column names, and then writes the table to the results block.

14.8.2.22 `:rownames`

The `:rownames` header argument can take on values `yes` or `no` values. The default is `no`. Note that `emacs-lisp` code blocks ignore `:rownames` header argument because of the ease of table-handling in Emacs.

- `no` Org will not pre-process row names.
- `yes` If an input table has row names, then Org removes the row names, processes the table, puts back the row names, and then writes the table to the results block.

```
#+NAME: with-rownames
| one | 1 | 2 | 3 | 4 |  5 |
| two | 6 | 7 | 8 | 9 | 10 |

#+NAME: echo-table-once-again
#+BEGIN_SRC python :var tab=with-rownames :rownames yes
  return [[val + 10 for val in row] for row in tab]
#+END_SRC

#+RESULTS: echo-table-once-again
| one | 11 | 12 | 13 | 14 | 15 |
| two | 16 | 17 | 18 | 19 | 20 |
```

Note that row names have to accounted for when using variable indexing (see Section 14.8.2.1 [var], page 208) because row names are not removed for indexing.

14.8.2.23 `:shebang`

This header argument can turn results into executable script files. By setting the `:shebang` header argument to a string value (for example, `:shebang "#!/bin/bash"`), Org inserts that string as the first line of the tangled file that the '`src`' code block is extracted to. Org then turns on the tangled file's executable permission.

14.8.2.24 `:tangle-mode`

The `tangle-mode` header argument specifies what permissions to set for tangled files by `set-file-modes`. For example, to make read-only tangled file, use `:tangle-mode (identity #o444)`. To make it executable, use `:tangle-mode (identity #o755)`.

On 'src' code blocks with **shebang** (see Section 14.8.2.23 [shebang], page 221) header argument, Org will automatically set the tangled file to executable permissions. But this can be overridden with custom permissions using **tangle-mode** header argument.

When multiple 'src' code blocks tangle to a single file with different and conflicting **tangle-mode** header arguments, Org's behavior is undefined.

14.8.2.25 :eval

The **:eval** header argument can limit evaluation of specific code blocks. It is useful for protection against evaluating untrusted 'src' code blocks by prompting for a confirmation. This protection is independent of the **org-confirm-babel-evaluate** setting.

`never` or `no`
: Org will never evaluate this 'src' code block.

`query`
: Org prompts the user for permission to evaluate this 'src' code block.

`never-export` or `no-export`
: Org will not evaluate this 'src' code block when exporting, yet the user can evaluate this source block interactively.

`query-export`
: Org prompts the user for permission to export this 'src' code block.

If **:eval** header argument is not set for a source block, then Org determines whether to evaluate from the **org-confirm-babel-evaluate** variable (see Section 15.4 [Code evaluation security], page 229).

14.8.2.26 :wrap

The **:wrap** header argument marks the results block by appending strings to **#+BEGIN_** and **#+END_**. If no string is specified, Org wraps the results in a **#+BEGIN/END_RESULTS** block.

14.8.2.27 :post

The **:post** header argument is for post-processing results from 'src' block evaluation. When **:post** has any value, Org binds the results to ***this*** variable for easy passing to Section 14.8.2.1 [var], page 208 header argument specifications. That makes results available to other 'src' code blocks, or for even direct Emacs Lisp code execution.

The following two examples illustrate **:post** header argument in action. The first one shows how to attach **#+ATTR_LATEX:** line using **:post**.

```
#+name: attr_wrap
#+begin_src sh :var data="" :var width="\\textwidth" :results output
  echo "#+ATTR_LATEX: :width $width"
  echo "$data"
#+end_src

#+header: :file /tmp/it.png
#+begin_src dot :post attr_wrap(width="5cm", data=*this*) :results drawer
  digraph{
          a -> b;
          b -> c;
```

```
        c -> a;
    }
#+end_src

#+RESULTS:
:RESULTS:
#+ATTR_LATEX :width 5cm
[[file:/tmp/it.png]]
:END:
```

The second example shows use of `:colnames` in `:post` to pass data between 'src' code blocks.

```
#+name: round-tbl
#+begin_src emacs-lisp :var tbl="" fmt="%.3f"
  (mapcar (lambda (row)
            (mapcar (lambda (cell)
                      (if (numberp cell)
                          (format fmt cell)
                        cell))
                    row))
          tbl)
#+end_src

#+begin_src R :colnames yes :post round-tbl[:colnames yes](*this*)
set.seed(42)
data.frame(foo=rnorm(1))
#+end_src

#+RESULTS:
|   foo |
|-------|
| 1.371 |
```

14.8.2.28 :prologue

The `prologue` header argument is for appending to the top of the code block for execution. For example, a clear or reset code at the start of new execution of a 'src' code block. A `reset` for 'gnuplot': `:prologue "reset"`. See also Section 14.8.2.29 [epilogue], page 223.

```
(add-to-list 'org-babel-default-header-args:gnuplot
             '((:prologue . "reset")))
```

14.8.2.29 :epilogue

The value of the `epilogue` header argument is for appending to the end of the code block for execution. See also Section 14.8.2.28 [prologue], page 223.

14.9 Results of evaluation

How Org handles results of a code block execution depends on many header arguments working together. Here is only a summary of these. For an enumeration of all the header arguments that affect results, see Section 14.8.2.2 [results], page 212.

The primary determinant is the execution context. Is it in a `:session` or not? Orthogonal to that is if the expected result is a `:results value` or `:results output`, which is a concatenation of output from start to finish of the '`src`' code block's evaluation.

	Non-session	**Session**
`:results value`	value of last expression	value of last expression
`:results output`	contents of STDOUT	concatenation of interpreter output

For `:session` and non-session, the `:results value` turns the results into an Org mode table format. Single values are wrapped in a one dimensional vector. Rows and columns of a table are wrapped in a two-dimensional vector.

14.9.1 Non-session

14.9.1.1 `:results value`

Default. Org gets the value by wrapping the code in a function definition in the language of the '`src`' block. That is why when using `:results value`, code should execute like a function and return a value. For languages like Python, an explicit `return` statement is mandatory when using `:results value`.

This is one of four evaluation contexts where Org automatically wraps the code in a function definition.

14.9.1.2 `:results output`

For `:results output`, the code is passed to an external process running the interpreter. Org returns the contents of the standard output stream as as text results.

14.9.2 Session

14.9.2.1 `:results value`

For `:results value` from a `:session`, Org passes the code to an interpreter running as an interactive Emacs inferior process. So only languages that provide interactive evaluation can have session support. Not all languages provide this support, such as '`C`' and '`ditaa`'. Even those that do support, such as '`Python`' and '`Haskell`', they impose limitations on allowable language constructs that can run interactively. Org inherits those limitations for those '`src`' code blocks running in a `:session`.

Org gets the value from the source code interpreter's last statement output. Org has to use language-specific methods to obtain the value. For example, from the variable `_` in '`Python`' and '`Ruby`', and the value of `.Last.value` in '`R`').

14.9.2.2 `:results output`

For `:results output`, Org passes the code to the interpreter running as an interactive Emacs inferior process. Org concatenates whatever text output emitted by the interpreter to return the collection as a result. Note that this collection is not the same as collected from STDOUT of a non-interactive interpreter running as an external process. Compare for example these two blocks:

```
#+BEGIN_SRC python :results output
 print "hello"
 2
 print "bye"
#+END_SRC

#+RESULTS:
: hello
: bye
```

In the above non-session mode, the "2" is not printed; so does not appear in results.

```
#+BEGIN_SRC python :results output :session
 print "hello"
 2
 print "bye"
#+END_SRC

#+RESULTS:
: hello
: 2
: bye
```

In the above :session mode, the interactive interpreter receives and prints "2". Results show that.

14.10 Noweb reference syntax

Org supports named blocks in Noweb style syntax. For Noweb literate programming details, see http://www.cs.tufts.edu/~nr/noweb/).

```
<<code-block-name>>
```

For the header argument :noweb yes, Org expands Noweb style references in the 'src' code block before evaluation.

For the header argument :noweb no, Org does not expand Noweb style references in the 'src' code block before evaluation.

The default is :noweb no. Org defaults to :noweb no so as not to cause errors in languages where Noweb syntax is ambiguous. Change Org's default to :noweb yes for languages where there is no risk of confusion.

Org offers a more flexible way to resolve Noweb style references (see Section 14.8.2.16 [noweb-ref], page 217).

Org can include the *results* of a code block rather than its body. To that effect, append parentheses, possibly including arguments, to the code block name, as show below.

```
<<code-block-name(optional arguments)>>
```

Note that when using the above approach to a code block's results, the code block name set by #+NAME keyword is required; the reference set by :noweb-ref will not work.

Here is an example that demonstrates how the exported content changes when Noweb style references are used with parentheses versus without.

With:

```
#+NAME: some-code
#+BEGIN_SRC python :var num=0 :results output :exports none
print(num*10)
#+END_SRC
```

this code block:

```
#+BEGIN_SRC text :noweb yes
<<some-code>>
#+END_SRC
```

expands to:

```
print(num*10)
```

Below, a similar Noweb style reference is used, but with parentheses, while setting a variable `num` to 10:

```
#+BEGIN_SRC text :noweb yes
<<some-code(num=10)>>
#+END_SRC
```

Note that now the expansion contains the *results* of the code block `some-code`, not the code block itself:

```
100
```

For faster tangling of large Org mode files, set `org-babel-use-quick-and-dirty-noweb-expansion` variable to `t`. The speedup comes at the expense of not correctly resolving inherited values of the `:noweb-ref` header argument.

14.11 Key bindings and useful functions

Many common Org mode key sequences are re-bound depending on the context.

Active key bindings in code blocks:

C-c C-c	`org-babel-execute-src-block`
C-c C-o	`org-babel-open-src-block-result`
M-up	`org-babel-load-in-session`
M-down	`org-babel-switch-to-session`

Active key bindings in Org mode buffer:

C-c C-v p or *C-c C-v C-p*	`org-babel-previous-src-block`
C-c C-v n or *C-c C-v C-n*	`org-babel-next-src-block`
C-c C-v e or *C-c C-v C-e*	`org-babel-execute-maybe`
C-c C-v o or *C-c C-v C-o*	`org-babel-open-src-block-result`
C-c C-v v or *C-c C-v C-v*	`org-babel-expand-src-block`
C-c C-v u or *C-c C-v C-u*	`org-babel-goto-src-block-head`
C-c C-v g or *C-c C-v C-g*	`org-babel-goto-named-src-block`
C-c C-v r or *C-c C-v C-r*	`org-babel-goto-named-result`

C-c C-v b or *C-c C-v C-b*	`org-babel-execute-buffer`
C-c C-v s or *C-c C-v C-s*	`org-babel-execute-subtree`
C-c C-v d or *C-c C-v C-d*	`org-babel-demarcate-block`
C-c C-v t or *C-c C-v C-t*	`org-babel-tangle`
C-c C-v f or *C-c C-v C-f*	`org-babel-tangle-file`
C-c C-v c or *C-c C-v C-c*	`org-babel-check-src-block`
C-c C-v j or *C-c C-v C-j*	`org-babel-insert-header-arg`
C-c C-v l or *C-c C-v C-l*	`org-babel-load-in-session`
C-c C-v i or *C-c C-v C-i*	`org-babel-lob-ingest`
C-c C-v I or *C-c C-v C-I*	`org-babel-view-src-block-info`
C-c C-v z or *C-c C-v C-z*	`org-babel-switch-to-session-with-code`
C-c C-v a or *C-c C-v C-a*	`org-babel-sha1-hash`
C-c C-v h or *C-c C-v C-h*	`org-babel-describe-bindings`
C-c C-v x or *C-c C-v C-x*	`org-babel-do-key-sequence-in-edit-buffer`

14.12 Batch execution

Org mode features, including working with source code facilities can be invoked from the command line. This enables building shell scripts for batch processing, running automated system tasks, and expanding Org mode's usefulness.

The sample script shows batch processing of multiple files using `org-babel-tangle`.

```
#!/bin/sh
# tangle files with org-mode
#
emacs -Q --batch --eval "
    (progn
      (require 'ob-tangle)
      (dolist (file command-line-args-left)
        (with-current-buffer (find-file-noselect file)
          (org-babel-tangle))))
  " "$@"
```

15 Miscellaneous

15.1 Completion

Org has in-buffer completions. Unlike minibuffer completions, which are useful for quick command interactions, Org's in-buffer completions are more suitable for content creation in Org documents. Type one or more letters and invoke the hot key to complete the text in-place. Depending on the context and the keys, Org will offer different types of completions. No minibuffer is involved. Such mode-specific hot keys have become an integral part of Emacs and Org provides several shortcuts.

M-TAB Complete word at point

- At the beginning of a headline, complete TODO keywords.
- After '\', complete $\TeX$ symbols supported by the exporter.
- After '*', complete headlines in the current buffer so that they can be used in search links like '[[*find this headline]]'.
- After ':' in a headline, complete tags. The list of tags is taken from the variable org-tag-alist (possibly set through the '#+TAGS' in-buffer option, see Section 6.2 [Setting tags], page 59), or it is created dynamically from all tags used in the current buffer.
- After ':' and not in a headline, complete property keys. The list of keys is constructed dynamically from all keys used in the current buffer.
- After '[', complete link abbreviations (see Section 4.6 [Link abbreviations], page 44).
- After '#+', complete the special keywords like 'TYP_TODO' or file-specific 'OPTIONS'. After option keyword is complete, pressing *M-TAB* again will insert example settings for that option.
- After '#+STARTUP: ', complete startup keywords.
- When the point is anywhere else, complete dictionary words using Ispell.

If your desktop intercepts the combo *M-TAB* to switch windows, use *C-M-i* or *ESC TAB* as an alternative or customize your environment.

15.2 Easy templates

With just a few keystrokes, Org's easy templates inserts empty pairs of structural elements, such as #+BEGIN_SRC and #+END_SRC. Easy templates use an expansion mechanism, which is native to Org, in a process similar to yasnippet and other Emacs template expansion packages.

< *s* TAB expands to a 'src' code block.

< *l* TAB expands to:

#+BEGIN_EXPORT latex

#+END_EXPORT

Org comes with these pre-defined easy templates:

s #+BEGIN_SRC ... #+END_SRC

e	`#+BEGIN_EXAMPLE ... #+END_EXAMPLE`
q	`#+BEGIN_QUOTE ... #+END_QUOTE`
v	`#+BEGIN_VERSE ... #+END_VERSE`
c	`#+BEGIN_CENTER ... #+END_CENTER`
C	`#+BEGIN_COMMENT ... #+END_COMMENT`
l	`#+BEGIN_EXPORT latex ... #+END_EXPORT`
L	`#+LATEX:`
h	`#+BEGIN_EXPORT html ... #+END_EXPORT`
H	`#+HTML:`
a	`#+BEGIN_EXPORT ascii ... #+END_EXPORT`
A	`#+ASCII:`
i	`#+INDEX:` line
I	`#+INCLUDE:` line

More templates can added by customizing the variable `org-structure-template-alist`, whose docstring has additional details.

15.3 Speed keys

Single keystrokes can execute custom commands in an Org file when the cursor is on a headline. Without the extra burden of a meta or modifier key, Speed Keys can speed navigation or execute custom commands. Besides faster navigation, Speed Keys may come in handy on small mobile devices that do not have full keyboards. Speed Keys may also work on TTY devices known for their problems when entering Emacs keychords.

By default, Org has Speed Keys disabled. To activate Speed Keys, set the variable `org-use-speed-commands` to a non-`nil` value. To trigger a Speed Key, the cursor must be at the beginning of an Org headline, before any of the stars.

Org comes with a pre-defined list of Speed Keys. To add or modify Speed Keys, customize the variable, `org-speed-commands-user`. For more details, see the variable's docstring. With Speed Keys activated, *M-x org-speed-command-help*, or *?* when cursor is at the beginning of an Org headline, shows currently active Speed Keys, including the user-defined ones.

15.4 Code evaluation and security issues

Unlike plain text, running code comes with risk. Each '`src`' code block, in terms of risk, is equivalent to an executable file. Org therefore puts a few confirmation prompts by default. This is to alert the casual user from accidentally running untrusted code.

For users who do not run code blocks or write code regularly, Org's default settings should suffice. However, some users may want to tweak the prompts for fewer interruptions. To weigh the risks of automatic execution of code blocks, here are some details about code evaluation.

Org evaluates code in the following circumstances:

Source code blocks

Org evaluates '`src`' code blocks in an Org file during export. Org also evaluates a '`src`' code block with the *C-c C-c* key chord. Users exporting or running code blocks must load files only from trusted sources. Be wary of customizing variables that remove or alter default security measures.

org-confirm-babel-evaluate [User Option]
When t, Org prompts the user for confirmation before executing each code block. When nil, Org executes code blocks without prompting the user for confirmation. When this option is set to a custom function, Org invokes the function with these two arguments: the source code language and the body of the code block. The custom function must return either a t or nil, which determines if the user is prompted. Each source code language can be handled separately through this function argument.

For example, this function enables execution of 'ditaa' code +blocks without prompting:

```
(defun my-org-confirm-babel-evaluate (lang body)
  (not (string= lang "ditaa")))  ; don't ask for ditaa
(setq org-confirm-babel-evaluate 'my-org-confirm-babel-evaluate)
```

Following shell *and* elisp *links*
Org has two link types that can also directly evaluate code (see Section 4.3 [External links], page 39). Because such code is not visible, these links have a potential risk. Org therefore prompts the user when it encounters such links. The customization variables are:

org-confirm-shell-link-function [User Option]
Function that prompts the user before executing a shell link.

org-confirm-elisp-link-function [User Option]
Function that prompts the user before executing an Emacs Lisp link.

Formulas in tables
Org executes formulas in tables (see Section 3.5 [The spreadsheet], page 24) either through the *calc* or the *Emacs Lisp* interpreters.

15.5 Customization

Org has more than 500 variables for customization. They can be accessed through the usual *M-x org-customize RET* command. Or through the Org menu, Org->Customization->Browse Org Group. Org also has per-file settings for some variables (see Section 15.6 [In-buffer settings], page 230).

15.6 Summary of in-buffer settings

In-buffer settings start with '#+', followed by a keyword, a colon, and then a word for each setting. Org accepts multiple settings on the same line. Org also accepts multiple lines for a keyword. This manual describes these settings throughout. A summary follows here.

C-c C-c activates any changes to the in-buffer settings. Closing and reopening the Org file in Emacs also activates the changes.

#+ARCHIVE: %s_done::
Sets the archive location of the agenda file. This location applies to the lines until the next '#+ARCHIVE' line, if any, in the Org file. The first archive location in the Org file also applies to any entries before it. The corresponding variable is org-archive-location.

#+CATEGORY:
: Sets the category of the agenda file, which applies to the entire document.

#+COLUMNS: %25ITEM ...
: Sets the default format for columns view. Org uses this format for column views where there is no `COLUMNS` property.

#+CONSTANTS: name1=value1 ...
: Set file-local values for constants that table formulas can use. This line sets the local variable `org-table-formula-constants-local`. The global version of this variable is `org-table-formula-constants`.

#+FILETAGS: :tag1:tag2:tag3:
: Set tags that all entries in the file will inherit from here, including the top-level entries.

#+LINK: linkword replace
: Each line specifies one abbreviation for one link. Use multiple `#+LINK:` lines for more, see Section 4.6 [Link abbreviations], page 44. The corresponding variable is `org-link-abbrev-alist`.

#+PRIORITIES: highest lowest default
: This line sets the limits and the default for the priorities. All three must be either letters A–Z or numbers 0–9. The highest priority must have a lower ASCII number than the lowest priority.

#+PROPERTY: Property_Name Value
: This line sets a default inheritance value for entries in the current buffer, most useful for specifying the allowed values of a property.

#+SETUPFILE: file or URL
: The setup file or a URL pointing to such file is for additional in-buffer settings. Org loads this file and parses it for any settings in it only when Org opens the main file. If URL is specified, the contents are downloaded and stored in a temporary file cache. *C-c C-c* on the settings line will parse and load the file, and also reset the temporary file cache. Org also parses and loads the document during normal exporting process. Org parses the contents of this document as if it was included in the buffer. It can be another Org file. To visit the file (not a URL), *C-c '* while the cursor is on the line with the file name.

#+STARTUP:
: Startup options Org uses when first visiting a file.

  The first set of options deals with the initial visibility of the outline tree. The corresponding variable for global default settings is `org-startup-folded` with a default value of `t`, which is the same as `overview`.

`overview`	top-level headlines only
`content`	all headlines
`showall`	no folding of any entries
`showeverything`	show even drawer contents

  Dynamic virtual indentation is controlled by the variable `org-startup-indented`

`indent` start with `org-indent-mode` turned on
`noindent` start with `org-indent-mode` turned off

Aligns tables consistently upon visiting a file; useful for restoring narrowed table columns. The corresponding variable is `org-startup-align-all-tables` with `nil` as default value.

`align` align all tables
`noalign` donflt align tables on startup

Whether Org should automatically display inline images. The corresponding variable is `org-startup-with-inline-images`, with a default value `nil` to avoid delays when visiting a file.

`inlineimages` show inline images
`noinlineimages` donflt show inline images on startup

Whether Org should automatically convert LaTeX fragments to images. The variable `org-startup-with-latex-preview`, which controls this setting, is set to `nil` by default to avoid startup delays.

`latexpreview` preview LaTeX fragments
`nolatexpreview` donflt preview LaTeX fragments

Logging the closing and reopening of TODO items and clock intervals can be configured using these options (see variables `org-log-done`, `org-log-note-clock-out` and `org-log-repeat`)

`logdone` record a timestamp when an item is marked DONE
`lognotedone` record timestamp and a note when DONE
`nologdone` donflt record when items are marked DONE
`logrepeat` record a time when reinstating a repeating item
`lognoterepeat` record a note when reinstating a repeating item
`nologrepeat` do not record when reinstating repeating item
`lognoteclock-out` record a note when clocking out
`nolognoteclock-out` donflt record a note when clocking out
`logreschedule` record a timestamp when scheduling time changes
`lognotereschedule` record a note when scheduling time changes
`nologreschedule` do not record when a scheduling date changes
`logredeadline` record a timestamp when deadline changes
`lognoteredeadline` record a note when deadline changes
`nologredeadline` do not record when a deadline date changes
`logrefile` record a timestamp when refiling
`lognoterefile` record a note when refiling
`nologrefile` do not record when refiling
`logdrawer` store log into drawer
`nologdrawer` store log outside of drawer
`logstatesreversed` reverse the order of states notes
`nologstatesreversed` do not reverse the order of states notes

These options hide leading stars in outline headings, and indent outlines. The corresponding variables are `org-hide-leading-stars` and `org-odd-levels-only`, both with a default setting of `nil` (meaning `showstars` and `oddeven`).

`hidestars`	hide all stars on the headline except one.
`showstars`	show all stars on the headline
`indent`	virtual indents according to the outline level
`noindent`	no virtual indents
`odd`	show odd outline levels only (1,3,...)
`oddeven`	show all outline levels

To turn on custom format overlays over timestamps (variables `org-put-time-stamp-overlays` and `org-time-stamp-overlay-formats`), use

`customtime`	overlay custom time format

The following options influence the table spreadsheet (variable `constants-unit-system`).

`constcgs`	`constants.el` should use the c-g-s unit system
`constSI`	`constants.el` should use the SI unit system

For footnote settings, use the following keywords. The corresponding variables are `org-footnote-define-inline`, `org-footnote-auto-label`, and `org-footnote-auto-adjust`.

`fninline`	define footnotes inline
`fnnoinline`	define footnotes in separate section
`fnlocal`	define footnotes near first reference, but not inline
`fnprompt`	prompt for footnote labels
`fnauto`	create `[fn:1]`-like labels automatically (default)
`fnconfirm`	offer automatic label for editing or confirmation
`fnplain`	create `[1]`-like labels automatically
`fnadjust`	automatically renumber and sort footnotes
`nofnadjust`	do not renumber and sort automatically

To hide blocks on startup, use these keywords. The corresponding variable is `org-hide-block-startup`.

`hideblocks`	Hide all begin/end blocks on startup
`nohideblocks`	Do not hide blocks on startup

The display of entities as UTF-8 characters is governed by the variable `org-pretty-entities` and the keywords

`entitiespretty`	Show entities as UTF-8 characters where possible
`entitiesplain`	Leave entities plain

#+TAGS: TAG1(c1) TAG2(c2)

These lines specify valid tags for this file. Org accepts multiple tags lines. Tags could correspond to the *fast tag selection* keys. The corresponding variable is `org-tag-alist`.

#+TBLFM: This line is for formulas for the table directly above. A table can have multiple '`#+TBLFM:`' lines. On table recalculation, Org applies only the first '`#+TBLFM:`' line. For details see [Using multiple #+TBLFM lines], page 33 in Section 3.5.8 [Editing and debugging formulas], page 31.

#+TITLE:, #+AUTHOR:, #+EMAIL:, #+LANGUAGE:, #+DATE:,
#+OPTIONS:, #+BIND:,
#+SELECT_TAGS:, #+EXCLUDE_TAGS:
These lines provide settings for exporting files. For more details see Section 12.2 [Export settings], page 138.

#+TODO: #+SEQ_TODO: #+TYP_TODO:
These lines set the TODO keywords and their significance to the current file. The corresponding variable is org-todo-keywords.

15.7 The very busy C-c C-c key

The *C-c C-c* key in Org serves many purposes depending on the context. It is probably the most over-worked, multi-purpose key combination in Org. Its uses are well-documented through out this manual, but here is a consolidated list for easy reference.

- If any highlights shown in the buffer from the creation of a sparse tree, or from clock display, remove such highlights.
- If the cursor is in one of the special #+KEYWORD lines, scan the buffer for these lines and update the information. Also reset the Org file cache used to temporary store the contents of URLs used as values for keywords like #+SETUPFILE.
- If the cursor is inside a table, realign the table. The table realigns even if automatic table editor is turned off.
- If the cursor is on a #+TBLFM line, re-apply the formulas to the entire table.
- If the current buffer is a capture buffer, close the note and file it. With a prefix argument, also jump to the target location after saving the note.
- If the cursor is on a <<<target>>>, update radio targets and corresponding links in this buffer.
- If the cursor is on a property line or at the start or end of a property drawer, offer property commands.
- If the cursor is at a footnote reference, go to the corresponding definition, and *vice versa*.
- If the cursor is on a statistics cookie, update it.
- If the cursor is in a plain list item with a checkbox, toggle the status of the checkbox.
- If the cursor is on a numbered item in a plain list, renumber the ordered list.
- If the cursor is on the #+BEGIN line of a dynamic block, the block is updated.
- If the cursor is at a timestamp, fix the day name in the timestamp.

15.8 A cleaner outline view

Org's default outline with stars and no indents can become too cluttered for short documents. For *book-like* long documents, the effect is not as noticeable. Org provides an alternate stars and indentation scheme, as shown on the right in the following table. It uses only one star and indents text to line with the heading:

```
* Top level headline              |    * Top level headline
** Second level                   |      * Second level
*** 3rd level                     |        * 3rd level
some text                         |          some text
*** 3rd level                     |        * 3rd level
more text                         |          more text
* Another top level headline      |    * Another top level headline
```

To turn this mode on, use the minor mode, `org-indent-mode`. Text lines that are not headlines are prefixed with spaces to vertically align with the headline text[1].

To make more horizontal space, the headlines are shifted by two stars. This can be configured by the `org-indent-indentation-per-level` variable. Only one star on each headline is visible, the rest are masked with the same font color as the background. This font face can be configured with the `org-hide` variable.

Note that turning on `org-indent-mode` sets `org-hide-leading-stars` to `t` and `org-adapt-indentation` to `nil`; '2.' below shows how this works.

To globally turn on `org-indent-mode` for all files, customize the variable `org-startup-indented`.

To turn on indenting for individual files, use `#+STARTUP` option as follows:

```
#+STARTUP: indent
```

Indent on startup makes Org use hard spaces to align text with headings as shown in examples below.

1. *Indentation of text below headlines*
   Indent text to align with the headline.

   ```
   *** 3rd level
       more text, now indented
   ```

   Org adapts indentations with paragraph filling, line wrapping, and structure editing[2].
2. *Hiding leading stars*
   Org can make leading stars invisible. For global preference, configure the variable `org-hide-leading-stars`. For per-file preference, use these file `#+STARTUP` options:

   ```
   #+STARTUP: hidestars
   #+STARTUP: showstars
   ```

   With stars hidden, the tree is shown as:

   ```
   * Top level headline
    * Second level
     * 3rd level
     ...
   ```

   Because Org makes the font color same as the background color to hide to stars, sometimes `org-hide` face may need tweaking to get the effect right. For some black and white combinations, `grey90` on a white background might mask the stars better.

[1] The `org-indent-mode` also sets the `wrap-prefix` correctly for indenting and wrapping long lines of headlines or text. This minor mode handles `visual-line-mode` and directly applied settings through `word-wrap`.

[2] Also see the variable `org-adapt-indentation`.

3. Using stars for only odd levels, 1, 3, 5, ..., can also clean up the clutter. This removes two stars from each level[3]. For Org to properly handle this cleaner structure during edits and exports, configure the variable `org-odd-levels-only`. To set this per-file, use either one of the following lines:

```
#+STARTUP: odd
#+STARTUP: oddeven
```

   To switch between single and double stars layouts, use *M-x org-convert-to-odd-levels RET* and *M-x org-convert-to-oddeven-levels*.

15.9 Using Org on a tty

Org provides alternative key bindings for TTY and modern mobile devices that cannot handle cursor keys and complex modifier key chords. Some of these workarounds may be more cumbersome than necessary. Users should look into customizing these further based on their usage needs. For example, the normal *S-cursor* for editing timestamp might be better with *C-c .* chord.

Default	Alternative 1	Speed key	Alternative 2
S-TAB	*C-u TAB*	*C*	
M-left	*C-c C-x l*	*l*	*Esc left*
M-S-left	*C-c C-x L*	*L*	
M-right	*C-c C-x r*	*r*	*Esc right*
M-S-right	*C-c C-x R*	*R*	
M-up	*C-c C-x u*		*Esc up*
M-S-up	*C-c C-x U*	*U*	
M-down	*C-c C-x d*		*Esc down*
M-S-down	*C-c C-x D*	*D*	
S-RET	*C-c C-x c*		
M-RET	*C-c C-x m*		*Esc RET*
M-S-RET	*C-c C-x M*		
S-left	*C-c left*		
S-right	*C-c right*		
S-up	*C-c up*		
S-down	*C-c down*		
C-S-left	*C-c C-x left*		
C-S-right	*C-c C-x right*		

15.10 Interaction with other packages

Org's compatibility and the level of interaction with other Emacs packages are documented here.

---

[3] Because '`LEVEL=2`' has 3 stars, '`LEVEL=3`' has 4 stars, and so on

15.10.1 Packages that Org cooperates with

`calc.el` by Dave Gillespie
: Org uses the Calc package for tables to implement spreadsheet functionality (see Section 3.5 [The spreadsheet], page 24). Org also uses Calc for embedded calculations. See Section "Embedded Mode" in *GNU Emacs Calc Manual*.

`constants.el` by Carsten Dominik
: Org can use names for constants in formulas in tables. Org can also use calculation suffixes for units, such as 'M' for 'Mega'. For a standard collection of such constants, install the `constants` package. Install version 2.0 of this package, available at `https://staff.fnwi.uva.nl/c.dominik/Tools/`. Org checks if the function `constants-get` has been autoloaded. Installation instructions are in the file, `constants.el`.

`cdlatex.el` by Carsten Dominik
: Org mode can use CDLaTeX package to efficiently enter LaTeX fragments into Org files (see Section 11.8.3 [CDLaTeX mode], page 136).

`imenu.el` by Ake Stenhoff and Lars Lindberg
: Imenu creates dynamic menus based on an index of items in a file. Org mode supports Imenu menus. Enable it with a mode hook as follows:

```
(add-hook 'org-mode-hook
          (lambda () (imenu-add-to-menubar "Imenu")))
```

By default the Imenu index is two levels deep. Change the index depth using thes variable, `org-imenu-depth`.

`speedbar.el` by Eric M. Ludlam
: Speedbar package creates a special Emacs frame for displaying files and index items in files. Org mode supports Speedbar; users can drill into Org files directly from the Speedbar. The `<` in the Speedbar frame tweaks the agenda commands to that file or to a subtree.

`table.el` by Takaaki Ota
: Complex ASCII tables with automatic line wrapping, column- and row-spanning, and alignment can be created using the Emacs table package by Takaaki Ota. Org mode recognizes such tables and export them properly. *C-c '* to edit these tables in a special buffer, much like Org's '`src`' code blocks. Because of interference with other Org mode functionality, Takaaki Ota tables cannot be edited directly in the Org buffer.

C-c ' `org-edit-special`
: Edit a `table.el` table. Works when the cursor is in a table.el table.

C-c ~ `org-table-create-with-table.el`
: Insert a `table.el` table. If there is already a table at point, this command converts it between the `table.el` format and the Org mode format. See the documentation string of the command `org-convert-table` for details.

15.10.2 Packages that conflict with Org mode

In Emacs, `shift-selection-mode` combines cursor motions with shift key to enlarge regions. Emacs sets this mode by default. This conflicts with Org's use of *S-cursor* commands to change timestamps, TODO keywords, priorities, and item bullet types, etc. Since *S-cursor* commands outside of specific contexts don't do anything, Org offers the variable `org-support-shift-select` for customization. Org mode accommodates shift selection by (i) making it available outside of the special contexts where special commands apply, and (ii) extending an existing active region even if the cursor moves across a special context.

`CUA.el` by Kim. F. Storm
: Org key bindings conflict with *S-<cursor>* keys used by CUA mode. For Org to relinquish these bindings to CUA mode, configure the variable `org-replace-disputed-keys`. When set, Org moves the following key bindings in Org files, and in the agenda buffer (but not during date selection).

S-UP	⇒	M-p	S-DOWN	⇒	M-n
S-LEFT	⇒	M--	S-RIGHT	⇒	M-+
C-S-LEFT	⇒	M-S--	C-S-RIGHT	⇒	M-S-+

Yes, these are unfortunately more difficult to remember. To define a different replacement keys, look at the variable `org-disputed-keys`.

`ecomplete.el` by Lars Magne Ingebrigtsen larsi@gnus.org
: Ecomplete provides "electric" address completion in address header lines in message buffers. Sadly Orgtbl mode cuts ecompletes power supply: No completion happens when Orgtbl mode is enabled in message buffers while entering text in address header lines. If one wants to use ecomplete one should *not* follow the advice to automagically turn on Orgtbl mode in message buffers (see Section 3.4 [Orgtbl mode], page 24), but instead—after filling in the message headers—turn on Orgtbl mode manually when needed in the messages body.

`filladapt.el` by Kyle Jones
: Org mode tries to do the right thing when filling paragraphs, list items and other elements. Many users reported problems using both `filladapt.el` and Org mode, so a safe thing to do is to disable filladapt like this:

```
(add-hook 'org-mode-hook 'turn-off-filladapt-mode)
```

`yasnippet.el`
: The way Org mode binds the `TAB` key (binding to `[tab]` instead of `"\t"`) overrules YASnippet's access to this key. The following code fixed this problem:

```
(add-hook 'org-mode-hook
          (lambda ()
            (setq-local yas/trigger-key [tab])
            (define-key yas/keymap [tab] 'yas/next-field-or-maybe-expand)))
```

The latest version of yasnippet doesn't play well with Org mode. If the above code does not fix the conflict, first define the following function:

```
(defun yas/org-very-safe-expand ()
```

```
(let ((yas/fallback-behavior 'return-nil)) (yas/expand)))
```

Then tell Org mode to use that function:

```
(add-hook 'org-mode-hook
          (lambda ()
            (make-variable-buffer-local 'yas/trigger-key)
            (setq yas/trigger-key [tab])
            (add-to-list 'org-tab-first-hook 'yas/org-very-safe-e
            (define-key yas/keymap [tab] 'yas/next-field)))
```

`windmove.el` by Hovav Shacham

This package also uses the *S-<cursor>* keys, so everything written in the paragraph above about CUA mode also applies here. If you want make the windmove function active in locations where Org mode does not have special functionality on *S-cursor*, add this to your configuration:

```
;; Make windmove work in org-mode:
(add-hook 'org-shiftup-final-hook 'windmove-up)
(add-hook 'org-shiftleft-final-hook 'windmove-left)
(add-hook 'org-shiftdown-final-hook 'windmove-down)
(add-hook 'org-shiftright-final-hook 'windmove-right)
```

`viper.el` by Michael Kifer

Viper uses *C-c /* and therefore makes this key not access the corresponding Org mode command `org-sparse-tree`. You need to find another key for this command, or override the key in `viper-vi-global-user-map` with

```
(define-key viper-vi-global-user-map "C-c /" 'org-sparse-tree)
```

15.11 org-crypt.el

Org crypt encrypts the text of an Org entry, but not the headline, or properties. Org crypt uses the Emacs EasyPG library to encrypt and decrypt.

Any text below a headline that has a '`:crypt:`' tag will be automatically be encrypted when the file is saved. To use a different tag, customize the `org-crypt-tag-matcher` variable.

Suggested Org crypt settings in Emacs init file:

```
(require 'org-crypt)
(org-crypt-use-before-save-magic)
(setq org-tags-exclude-from-inheritance (quote ("crypt")))

(setq org-crypt-key nil)
  ;; GPG key to use for encryption
  ;; Either the Key ID or set to nil to use symmetric encryption.

(setq auto-save-default nil)
  ;; Auto-saving does not cooperate with org-crypt.el: so you need
  ;; to turn it off if you plan to use org-crypt.el quite often.
  ;; Otherwise, you'll get an (annoying) message each time you
  ;; start Org.
```

```
;; To turn it off only locally, you can insert this:
;;
;; # -*- buffer-auto-save-file-name: nil; -*-
```

Excluding the crypt tag from inheritance prevents encrypting previously encrypted text.

Appendix A Hacking

This appendix covers some areas where users can extend the functionality of Org.

A.1 Hooks

Org has a large number of hook variables for adding functionality. This appendix illustrates using a few. A complete list of hooks with documentation is maintained by the Worg project at http://orgmode.org/worg/doc.html#hooks.

A.2 Add-on packages

Various authors wrote a large number of add-on packages for Org.

These packages are not part of Emacs, but they are distributed as contributed packages with the separate release available at http://orgmode.org. See the contrib/README file in the source code directory for a list of contributed files. Worg page with more information is at: http://orgmode.org/worg/org-contrib/.

A.3 Adding hyperlink types

Org has many built-in hyperlink types (see Chapter 4 [Hyperlinks], page 38), and an interface for adding new link types. The example file, org-man.el, shows the process of adding Org links to Unix man pages, which look like this: '[[man:printf][The printf manpage]]':

```
;;; org-man.el - Support for links to manpages in Org

(require 'org)

(org-add-link-type "man" 'org-man-open)
(add-hook 'org-store-link-functions 'org-man-store-link)

(defcustom org-man-command 'man
  "The Emacs command to be used to display a man page."
  :group 'org-link
  :type '(choice (const man) (const woman)))

(defun org-man-open (path)
  "Visit the manpage on PATH.
PATH should be a topic that can be thrown at the man command."
  (funcall org-man-command path))

(defun org-man-store-link ()
  "Store a link to a manpage."
  (when (memq major-mode '(Man-mode woman-mode))
    ;; This is a man page, we do make this link
    (let* ((page (org-man-get-page-name))
           (link (concat "man:" page))
           (description (format "Manpage for %s" page)))
```

```
          (org-store-link-props
           :type "man"
           :link link
           :description description))))

(defun org-man-get-page-name ()
  "Extract the page name from the buffer name."
  ;; This works for both `Man-mode' and `woman-mode'.
  (if (string-match " \\(\\S-+\\)\\*" (buffer-name))
      (match-string 1 (buffer-name))
    (error "Cannot create link to this man page")))

(provide 'org-man)

;;; org-man.el ends here
```

To activate links to man pages in Org, enter this in the init file:

```
(require 'org-man)
```

A review of `org-man.el`:

1. First, `(require 'org)` ensures `org.el` is loaded.
2. The `org-add-link-type` defines a new link type with '`man`' prefix. The call contains the function to call that follows the link type.
3. The next line adds a function to `org-store-link-functions` that records a useful link with the command *`C-c l`* in a buffer displaying a man page.

The rest of the file defines necessary variables and functions. First is the customization variable `org-man-command`. It has two options, `man` and `woman`. Next is a function whose argument is the link path, which for man pages is the topic of the man command. To follow the link, the function calls the `org-man-command` to display the man page.

`C-c l` constructs and stores the link.

`C-c l` calls the function `org-man-store-link`, which first checks if the `major-mode` is appropriate. If check fails, the function returns `nil`. Otherwise the function makes a link string by combining the '`man:`' prefix with the man topic. The function then calls `org-store-link-props` with `:type` and `:link` properties. A `:description` property is an optional string that is displayed when the function inserts the link in the Org buffer.

`C-c C-l` inserts the stored link.

To define new link types, define a function that implements completion support with *`C-c C-l`*. This function should not accept any arguments but return the appropriate prefix and complete link string.

A.4 Adding export back-ends

Org's export engine makes it easy for writing new back-ends. The framework on which the engine was built makes it easy to derive new back-ends from existing ones.

The two main entry points to the export engine are: `org-export-define-backend` and `org-export-define-derived-backend`. To grok these functions, see `ox-latex.el` for an example of defining a new back-end from scratch, and `ox-beamer.el` for an example of deriving from an existing engine.

For creating a new back-end from scratch, first set its name as a symbol in an alist consisting of elements and export functions. To make the back-end visible to the export dispatcher, set `:menu-entry` keyword. For export options specific to this back-end, set the `:options-alist`.

For creating a new back-end from an existing one, set `:translate-alist` to an alist of export functions. This alist replaces the parent back-end functions.

For complete documentation, see the Org Export Reference on Worg.

A.5 Context-sensitive commands

Org has facilities for building context sensitive commands. Authors of Org add-ons can tap into this functionality.

Some Org commands change depending on the context. The most important example of this behavior is the *C-c C-c* (see Section 15.7 [The very busy C-c C-c key], page 234). Other examples are *M-cursor* and *M-S-cursor*.

These context sensitive commands work by providing a function that detects special context for that add-on and executes functionality appropriate for that context.

A.6 Tables and lists in arbitrary syntax

Because of Org's success in handling tables with Orgtbl, a frequently asked feature is to Org's usability functions to other table formats native to other modem's, such as LaTeX. This would be hard to do in a general way without complicated customization nightmares. Moreover, that would take Org away from its simplicity roots that Orgtbl has proven. There is, however, an alternate approach to accomplishing the same.

This approach involves implementing a custom *translate* function that operates on a native Org *source table* to produce a table in another format. This strategy would keep the excellently working Orgtbl simple and isolate complications, if any, confined to the translate function. To add more alien table formats, we just add more translate functions. Also the burden of developing custom translate functions for new table formats will be in the hands of those who know those formats best.

For an example of how this strategy works, see Orgstruct mode. In that mode, Bastien added the ability to use Org's facilities to edit and re-structure lists. He did by turning `orgstruct-mode` on, and then exporting the list locally to another format, such as HTML, LaTeX or Texinfo.

A.6.1 Radio tables

Radio tables are target locations for translated tables that are not near their source. Org finds the target location and inserts the translated table.

The key to finding the target location are the magic words `BEGIN/END RECEIVE ORGTBL`. They have to appear as comments in the current mode. If the mode is C, then:

```
/* BEGIN RECEIVE ORGTBL table_name */
/* END RECEIVE ORGTBL table_name */
```

At the location of source, Org needs a special line to direct Orgtbl to translate and to find the target for inserting the translated table. For example:

```
#+ORGTBL: SEND table_name translation_function arguments...
```

`table_name` is the table's reference name, which is also used in the receiver lines, and the `translation_function` is the Lisp function that translates. This line, in addition, may also contain alternating key and value arguments at the end. The translation function gets these values as a property list. A few standard parameters are already recognized and acted upon before the translation function is called:

`:skip N` Skip the first N lines of the table. Hlines do count; include them if they are to be skipped.

`:skipcols (n1 n2 ...)`
List of columns to be skipped. First Org automatically discards columns with calculation marks and then sends the table to the translator function, which then skips columns as specified in '`skipcols`'.

To keep the source table intact in the buffer without being disturbed when the source file is compiled or otherwise being worked on, use one of these strategies:

- Place the table in a block comment. For example, in C mode you could wrap the table between '`/*`' and '`*/`' lines.
- Put the table after an '`END`' statement. For example '`\bye`' in TEX and '`\end{document}`' in LATEX.
- Comment and uncomment each line of the table during edits. The *M-x* ***orgtbl-toggle-comment*** *RET* command makes toggling easy.

A.6.2 A LATEX example of radio tables

To wrap a source table in LATEX, use the `comment` environment provided by `comment.sty`. To activate it, put `\usepackage{comment}` in the document header. Orgtbl mode inserts a radio table skeleton[1] with the command *M-x* ***orgtbl-insert-radio-table*** *RET*, which prompts for a table name. For example, if '`salesfigures`' is the name, the template inserts:

```
% BEGIN RECEIVE ORGTBL salesfigures
% END RECEIVE ORGTBL salesfigures
\begin{comment}
#+ORGTBL: SEND salesfigures orgtbl-to-latex
| | |
\end{comment}
```

The line `#+ORGTBL: SEND` tells Orgtbl mode to use the function `orgtbl-to-latex` to convert the table to LATEX format, then insert the table at the target (receive) location named `salesfigures`. Now the table is ready for data entry. It can even use spreadsheet features[2]:

---

[1] By default this works only for LATEX, HTML, and Texinfo. Configure the variable `orgtbl-radio-table-templates` to install templates for other export formats.

[2] If the '`#+TBLFM`' line contains an odd number of dollar characters, this may cause problems with font-lock in LATEX mode. As shown in the example you can fix this by adding an extra line inside the `comment` environment that is used to balance the dollar expressions. If you are using AUCTEX with the font-latex library, a much better solution is to add the `comment` environment to the variable `LaTeX-verbatim-environments`.

```
% BEGIN RECEIVE ORGTBL salesfigures
% END RECEIVE ORGTBL salesfigures
\begin{comment}
#+ORGTBL: SEND salesfigures orgtbl-to-latex
| Month | Days | Nr sold | per day |
|-------+------+---------+---------|
| Jan   |   23 |      55 |     2.4 |
| Feb   |   21 |      16 |     0.8 |
| March |   22 |     278 |    12.6 |
#+TBLFM: $4=$3/$2;%.1f
% $ (optional extra dollar to keep font-lock happy, see footnote)
\end{comment}
```

After editing, *C-c C-c* inserts translated table at the target location, between the two marker lines.

For hand-made custom tables, note that the translator needs to skip the first two lines of the source table. Also the command has to *splice* out the target table without the header and footer.

```
\begin{tabular}{lrrr}
Month & \multicolumn{1}{c}{Days} & Nr.\ sold & per day\\
% BEGIN RECEIVE ORGTBL salesfigures
% END RECEIVE ORGTBL salesfigures
\end{tabular}
%
\begin{comment}
#+ORGTBL: SEND salesfigures orgtbl-to-latex :splice t :skip 2
| Month | Days | Nr sold | per day |
|-------+------+---------+---------|
| Jan   |   23 |      55 |     2.4 |
| Feb   |   21 |      16 |     0.8 |
| March |   22 |     278 |    12.6 |
#+TBLFM: $4=$3/$2;%.1f
\end{comment}
```

The LaTeX translator function `orgtbl-to-latex` is already part of Orgtbl mode and uses `tabular` environment by default to typeset the table and mark the horizontal lines with `\hline`. For additional parameters to control output, see Section A.6.3 [Translator functions], page 246:

`:splice nil/t`
: When non-`nil`, returns only table body lines; not wrapped in tabular environment. Default is `nil`.

`:fmt fmt`
: Format to warp each field. It should contain `%s` for the original field value. For example, to wrap each field value in dollar symbol, you could use `:fmt "$%s$"`. Format can also wrap a property list with column numbers and formats, for example `:fmt (2 "$%s$" 4 "%s\\%%")`. In place of a string, a function of one argument can be used; the function must return a formatted string.

`:efmt efmt`

Format numbers as exponentials. The spec should have `%s` twice for inserting mantissa and exponent, for example `"%s\\times10^{%s}"`. This may also be a property list with column numbers and formats, for example `:efmt (2 "$%s\\times10^{%s}$" 4 "$%s\\cdot10^{%s}$")`. After `efmt` has been applied to a value, `fmt` will also be applied. Functions with two arguments can be supplied instead of strings. By default, no special formatting is applied.

A.6.3 Translator functions

Orgtbl mode has built-in translator functions: `orgtbl-to-csv` (comma-separated values), `orgtbl-to-tsv` (TAB-separated values), `orgtbl-to-latex`, `orgtbl-to-html`, `orgtbl-to-texinfo`, `orgtbl-to-unicode` and `orgtbl-to-orgtbl`. They use the generic translator, `orgtbl-to-generic`, which delegates translations to various export back-ends.

Properties passed to the function through the '`ORGTBL SEND`' line take precedence over properties defined inside the function. For example, this overrides the default LaTeX line endings, '`\\`', with '`\\[2mm]`':

```
#+ORGTBL: SEND test orgtbl-to-latex :lend " \\\\[2mm]"
```

For a new language translator, define a converter function. It can be a generic function, such as shown in this example. It marks a beginning and ending of a table with '`!BTBL!`' and '`!ETBL!`'; a beginning and ending of lines with '`!BL!`' and '`!EL!`'; and uses a TAB for a field separator:

```
(defun orgtbl-to-language (table params)
  "Convert the orgtbl-mode TABLE to language."
  (orgtbl-to-generic
   table
   (org-combine-plists
    '(:tstart "!BTBL!" :tend "!ETBL!" :lstart "!BL!" :lend "!EL!" :sep "\t")
    params)))
```

The documentation for the `orgtbl-to-generic` function shows a complete list of parameters, each of which can be passed through to `orgtbl-to-latex`, `orgtbl-to-texinfo`, and any other function using that generic function.

For complicated translations the generic translator function could be replaced by a custom translator function. Such a custom function must take two arguments and return a single string containing the formatted table. The first argument is the table whose lines are a list of fields or the symbol `hline`. The second argument is the property list consisting of parameters specified in the '`#+ORGTBL: SEND`' line. Please share your translator functions by posting them to the Org users mailing list, emacs-orgmode@gnu.org.

A.6.4 Radio lists

Call the `org-list-insert-radio-list` function to insert a radio list template in HTML, LaTeX, and Texinfo mode documents. Sending and receiving radio lists works is the same as for radio tables (see Section A.6.1 [Radio tables], page 243) except for these differences:

- Orgstruct mode must be active.
- Use `ORGLST` keyword instead of `ORGTBL`.
- *C-c C-c* works only on the first list item.

Built-in translators functions are: `org-list-to-latex`, `org-list-to-html` and `org-list-to-texinfo`. They use the `org-list-to-generic` translator function. See its documentation for parameters for accurate customizations of lists. Here is a LaTeX example:

```
% BEGIN RECEIVE ORGLST to-buy
% END RECEIVE ORGLST to-buy
\begin{comment}
#+ORGLST: SEND to-buy org-list-to-latex
- a new house
- a new computer
  + a new keyboard
  + a new mouse
- a new life
\end{comment}
```

C-c C-c on '`a new house`' inserts the translated LaTeX list in-between the BEGIN and END marker lines.

A.7 Dynamic blocks

Org supports *dynamic blocks* in Org documents. They are inserted with begin and end markers like any other '`src`' code block, but the contents are updated automatically by a user function. For example, *C-c C-x C-r* inserts a dynamic table that updates the work time (see Section 8.4 [Clocking work time], page 80).

Dynamic blocks can have names and function parameters. The syntax is similar to '`src`' code block specifications:

```
#+BEGIN: myblock :parameter1 value1 :parameter2 value2 ...

#+END:
```

These command update dynamic blocks:

C-c C-x C-u `org-dblock-update`
: Update dynamic block at point.

C-u C-c C-x C-u
: Update all dynamic blocks in the current file.

Before updating a dynamic block, Org removes content between the BEGIN and END markers. Org then reads the parameters on the BEGIN line for passing to the writer function. If the function expects to access the removed content, then Org expects an extra parameter, `:content`, on the BEGIN line.

To syntax for calling a writer function with a named block, `myblock` is: `org-dblock-write:myblock`. Parameters come from the BEGIN line.

The following is an example of a dynamic block and a block writer function that updates the time when the function was last run:

```
#+BEGIN: block-update-time :format "on %m/%d/%Y at %H:%M"

#+END:
```

The dynamic block's writer function:

```
(defun org-dblock-write:block-update-time (params)
  (let ((fmt (or (plist-get params :format) "%d. %m. %Y")))
    (insert "Last block update at: "
            (format-time-string fmt))))
```

To keep dynamic blocks up-to-date in an Org file, use the function, `org-update-all-dblocks` in hook, such as `before-save-hook`. The `org-update-all-dblocks` function does not run if the file is not in Org mode.

Dynamic blocks, like any other block, can be narrowed with `org-narrow-to-block`.

A.8 Special agenda views

Org provides a special hook to further limit items in agenda views: `agenda`, `agenda*`[3], `todo`, `alltodo`, `tags`, `tags-todo`, `tags-tree`. Specify a custom function that tests inclusion of every matched item in the view. This function can also skip as much as is needed.

For a global condition applicable to agenda views, use the `org-agenda-skip-function-global` variable. Org uses a global condition with `org-agenda-skip-function` for custom searching.

This example defines a function for a custom view showing TODO items with WAITING status. Manually this is a multi step search process, but with a custom view, this can be automated as follows:

The custom function searches the subtree for the WAITING tag and returns `nil` on match. Otherwise it gives the location from where the search continues.

```
(defun my-skip-unless-waiting ()
  "Skip trees that are not waiting"
  (let ((subtree-end (save-excursion (org-end-of-subtree t))))
    (if (re-search-forward ":waiting:" subtree-end t)
        nil          ; tag found, do not skip
      subtree-end))) ; tag not found, continue after end of subtree
```

To use this custom function in a custom agenda command:

```
(org-add-agenda-custom-command
 '("b" todo "PROJECT"
   ((org-agenda-skip-function 'my-skip-unless-waiting)
    (org-agenda-overriding-header "Projects waiting for something: "))))
```

Note that this also binds `org-agenda-overriding-header` to a more meaningful string suitable for the agenda view.

Search for entries with a limit set on levels for the custom search. This is a general approach to creating custom searches in Org. To include all levels, use '`LEVEL>0`'[4]. Then to selectively pick the matched entries, use `org-agenda-skip-function`, which also accepts Lisp forms, such as `org-agenda-skip-entry-if` and `org-agenda-skip-subtree-if`. For example:

---

[3] The `agenda*` view is the same as `agenda` except that it only considers *appointments*, i.e., scheduled and deadline items that have a time specification '`[h]h:mm`' in their time-stamps.

[4] Note that, for `org-odd-levels-only`, a level number corresponds to order in the hierarchy, not to the number of stars.

`(org-agenda-skip-entry-if 'scheduled)`
: Skip current entry if it has been scheduled.

`(org-agenda-skip-entry-if 'notscheduled)`
: Skip current entry if it has not been scheduled.

`(org-agenda-skip-entry-if 'deadline)`
: Skip current entry if it has a deadline.

`(org-agenda-skip-entry-if 'scheduled 'deadline)`
: Skip current entry if it has a deadline, or if it is scheduled.

`(org-agenda-skip-entry-if 'todo '("TODO" "WAITING"))`
: Skip current entry if the TODO keyword is TODO or WAITING.

`(org-agenda-skip-entry-if 'todo 'done)`
: Skip current entry if the TODO keyword marks a DONE state.

`(org-agenda-skip-entry-if 'timestamp)`
: Skip current entry if it has any timestamp, may also be deadline or scheduled.

`(org-agenda-skip-entry-if 'regexp "regular expression")`
: Skip current entry if the regular expression matches in the entry.

`(org-agenda-skip-entry-if 'notregexp "regular expression")`
: Skip current entry unless the regular expression matches.

`(org-agenda-skip-subtree-if 'regexp "regular expression")`
: Same as above, but check and skip the entire subtree.

The following is an example of a search for 'WAITING' without the special function:

```
(org-add-agenda-custom-command
 '("b" todo "PROJECT"
   ((org-agenda-skip-function '(org-agenda-skip-subtree-if
                               'regexp ":waiting:"))
    (org-agenda-overriding-header "Projects waiting for something: "))))
```

A.9 Speeding up your agendas

Some agenda commands slow down when the Org files grow in size or number. Here are tips to speed up:

1. Reduce the number of Org agenda files to avoid slowdowns due to hard drive accesses.
2. Reduce the number of 'DONE' and archived headlines so agenda operations that skip over these can finish faster.
3. Do not dim blocked tasks:
   ```
   (setq org-agenda-dim-blocked-tasks nil)
   ```
4. Stop preparing agenda buffers on startup:
   ```
   (setq org-agenda-inhibit-startup nil)
   ```
5. Disable tag inheritance for agendas:
   ```
   (setq org-agenda-use-tag-inheritance nil)
   ```

These options can be applied to selected agenda views. For more details about generation of agenda views, see the docstrings for the relevant variables, and this dedicated Worg page for agenda optimization.

A.10 Extracting agenda information

Org provides commands to access agendas through Emacs batch mode. Through this command-line interface, agendas are automated for further processing or printing.

`org-batch-agenda` creates an agenda view in ASCII and outputs to STDOUT. This command takes one string parameter. When string length=1, Org uses it as a key to `org-agenda-custom-commands`. These are the same ones available through *C-c* a.

This example command line directly prints the TODO list to the printer:

```
emacs -batch -l ~/.emacs -eval '(org-batch-agenda "t")' | lpr
```

When the string parameter length is two or more characters, Org matches it with tags/TODO strings. For example, this example command line prints items tagged with '`shop`', but excludes items tagged with '`NewYork`':

```
emacs -batch -l ~/.emacs                                      \
      -eval '(org-batch-agenda "+shop-NewYork")' | lpr
```

An example showing on-the-fly parameter modifications:

```
emacs -batch -l ~/.emacs                                      \
   -eval '(org-batch-agenda "a"                               \
            org-agenda-span (quote month)                     \
            org-agenda-include-diary nil                      \
            org-agenda-files (quote ("~/org/project.org")))'  \
   | lpr
```

which will produce an agenda for the next 30 days from just the `~/org/projects.org` file.

For structured processing of agenda output, use `org-batch-agenda-csv` with the following fields:

`category`	The category of the item
`head`	The headline, without TODO keyword, TAGS and PRIORITY
`type`	The type of the agenda entry, can be
	`todo` — selected in TODO match
	`tagsmatch` — selected in tags match
	`diary` — imported from diary
	`deadline` — a deadline
	`scheduled` — scheduled
	`timestamp` — appointment, selected by timestamp
	`closed` — entry was closed on date
	`upcoming-deadline` — warning about nearing deadline
	`past-scheduled` — forwarded scheduled item
	`block` — entry has date block including date
`todo`	The TODO keyword, if any
`tags`	All tags including inherited ones, separated by colons
`date`	The relevant date, like 2007-2-14
`time`	The time, like 15:00-16:50

`extra`	String with extra planning info
`priority-l`	The priority letter if any was given
`priority-n`	The computed numerical priority

If the selection of the agenda item was based on a timestamp, including those items with '`DEADLINE`' and '`SCHEDULED`' keywords, then Org includes date and time in the output.

If the selection of the agenda item was based on a timestamp (or deadline/scheduled), then Org includes date and time in the output.

Here is an example of a post-processing script in Perl. It takes the CSV output from Emacs and prints with a checkbox:

```
#!/usr/bin/perl

# define the Emacs command to run
$cmd = "emacs -batch -l ~/.emacs -eval '(org-batch-agenda-csv \"t\")'";

# run it and capture the output
$agenda = qx{$cmd 2>/dev/null};

# loop over all lines
foreach $line (split(/\n/,$agenda)) {
  # get the individual values
  ($category,$head,$type,$todo,$tags,$date,$time,$extra,
   $priority_l,$priority_n) = split(/,/,$line);
  # process and print
  print "[ ] $head\n";
}
```

A.11 Using the property API

Functions for working with properties.

`org-entry-properties` **&optional** *pom which* [Function]

Get all properties of the entry at point-or-marker POM.
This includes the TODO keyword, the tags, time strings for deadline, scheduled, and clocking, and any additional properties defined in the entry. The return value is an alist. Keys may occur multiple times if the property key was used several times.
POM may also be `nil`, in which case the current entry is used. If WHICH is `nil` or `all`, get all properties. If WHICH is `special` or `standard`, only get that subclass.

`org-entry-get` *pom property* **&optional** *inherit* [Function]

Get value of `PROPERTY` for entry at point-or-marker `POM`. By default, this only looks at properties defined locally in the entry. If `INHERIT` is non-`nil` and the entry does not have the property, then also check higher levels of the hierarchy. If `INHERIT` is the symbol `selective`, use inheritance if and only if the setting of `org-use-property-inheritance` selects `PROPERTY` for inheritance.

`org-entry-delete` *pom property* [Function]

Delete the property `PROPERTY` from entry at point-or-marker POM.

`org-entry-put` *pom property value* [Function]
Set `PROPERTY` to `VALUE` for entry at point-or-marker POM.

`org-buffer-property-keys` **&optional** *include-specials* [Function]
Get all property keys in the current buffer.

`org-insert-property-drawer` [Function]
Insert a property drawer for the current entry.

`org-entry-put-multivalued-property` *pom property* **&rest** *values* [Function]
Set `PROPERTY` at point-or-marker `POM` to `VALUES`. `VALUES` should be a list of strings. They will be concatenated, with spaces as separators.

`org-entry-get-multivalued-property` *pom property* [Function]
Treat the value of the property `PROPERTY` as a whitespace-separated list of values and return the values as a list of strings.

`org-entry-add-to-multivalued-property` *pom property value* [Function]
Treat the value of the property `PROPERTY` as a whitespace-separated list of values and make sure that `VALUE` is in this list.

`org-entry-remove-from-multivalued-property` *pom property value* [Function]
Treat the value of the property `PROPERTY` as a whitespace-separated list of values and make sure that `VALUE` is *not* in this list.

`org-entry-member-in-multivalued-property` *pom property value* [Function]
Treat the value of the property `PROPERTY` as a whitespace-separated list of values and check if `VALUE` is in this list.

`org-property-allowed-value-functions` [User Option]
Hook for functions supplying allowed values for a specific property. The functions must take a single argument, the name of the property, and return a flat list of allowed values. If '`:ETC`' is one of the values, use the values as completion help, but allow also other values to be entered. The functions must return `nil` if they are not responsible for this property.

A.12 Using the mapping API

Org has sophisticated mapping capabilities for finding entries. Org uses this functionality internally for generating agenda views. Org also exposes an API for executing arbitrary functions for each selected entry. The API's main entry point is:

`org-map-entries` *func* **&optional** *match scope* **&rest** *skip* [Function]
Call '`FUNC`' at each headline selected by `MATCH` in `SCOPE`.

'`FUNC`' is a function or a Lisp form. With the cursor positioned at the beginning of the headline, call the function without arguments. Org returns an alist of return values of calls to the function.

To avoid preserving point, Org wraps the call to `FUNC` in save-excursion form. After evaluation, Org moves the cursor to the end of the line that was just processed. Search continues from that point forward. This may not always work as expected under some conditions, such as if the current sub-tree was removed by a previous archiving operation. In such rare circumstances, Org skips the next entry entirely when it should not. To stop Org from such skips, make 'FUNC' set the variable `org-map-continue-from` to a specific buffer position.

'MATCH' is a tags/property/TODO match. Org iterates only matched headlines. Org iterates over all headlines when `MATCH` is `nil` or `t`.

'SCOPE' determines the scope of this command. It can be any of:

`nil` the current buffer, respecting the restriction if any

`tree` the subtree started with the entry at point

`region` The entries within the active region, if any

`file` the current buffer, without restriction

`file-with-archives`
: the current buffer, and any archives associated with it

`agenda` all agenda files

`agenda-with-archives`
: all agenda files with any archive files associated with them

`(file1 file2 ...)`
: if this is a list, all files in the list will be scanned

The remaining args are treated as settings for the scanner's skipping facilities. Valid args are:

`archive` skip trees with the archive tag

`comment` skip trees with the COMMENT keyword

`function or Lisp form`
: will be used as value for `org-agenda-skip-function`, so whenever the function returns t, FUNC will not be called for that entry and search will continue from the point where the function leaves it

The mapping routine can call any arbitrary function, even functions that change meta data or query the property API (see Section A.11 [Using the property API], page 251). Here are some handy functions:

`org-todo` **&optional** *arg* [Function]
: Change the TODO state of the entry. See the docstring of the functions for the many possible values for the argument `ARG`.

`org-priority` **&optional** *action* [Function]
: Change the priority of the entry. See the docstring of this function for the possible values for `ACTION`.

`org-toggle-tag` *tag* **&optional** *onoff* [Function]
: Toggle the tag `TAG` in the current entry. Setting `ONOFF` to either `on` or `off` will not toggle tag, but ensure that it is either on or off.

`org-promote` [Function]
: Promote the current entry.

org-demote [Function]
Demote the current entry.

This example turns all entries tagged with TOMORROW into TODO entries with keyword UPCOMING. Org ignores entries in comment trees and archive trees.

```
(org-map-entries
 '(org-todo "UPCOMING")
 "+TOMORROW" 'file 'archive 'comment)
```

The following example counts the number of entries with TODO keyword WAITING, in all agenda files.

```
(length (org-map-entries t "/+WAITING" 'agenda))
```

Appendix B MobileOrg

MobileOrg is a companion mobile app that runs on iOS and Android devices. MobileOrg enables offline-views and capture support for an Org mode system that is rooted on a "real" computer. MobileOrg can record changes to existing entries.

The iOS implementation for the *iPhone/iPod Touch/iPad* series of devices, was started by Richard Moreland and is now in the hands Sean Escriva. Android users should check out MobileOrg Android by Matt Jones. Though the two implementations are not identical, they offer similar features.

This appendix describes Org's support for agenda view formats compatible with MobileOrg. It also describes synchronizing changes, such as to notes, between MobileOrg and the computer.

To change tags and TODO states in MobileOrg, first customize the variables `org-todo-keywords` and `org-tag-alist`. These should cover all the important tags and TODO keywords, even if Org files use only some of them. Though MobileOrg has in-buffer settings, it understands TODO states *sets* (see Section 5.2.5 [Per-file keywords], page 49) and *mutually exclusive* tags (see Section 6.2 [Setting tags], page 59) only for those set in these variables.

B.1 Setting up the staging area

MobileOrg needs access to a file directory on a server to interact with Emacs. With a public server, consider encrypting the files. MobileOrg version 1.5 supports encryption for the iPhone. Org also requires `openssl` installed on the local computer. To turn on encryption, set the same password in MobileOrg and in Emacs. Set the password in the variable `org-mobile-use-encryption`[1]. Note that even after MobileOrg encrypts the file contents, the file names will remain visible on the file systems of the local computer, the server, and the mobile device.

For a server to host files, consider options like Dropbox.com account[2]. On first connection, MobileOrg creates a directory `MobileOrg/` on Dropbox. Pass its location to Emacs through an init file variable as follows:

```
(setq org-mobile-directory "~/Dropbox/MobileOrg")
```

Org copies files to the above directory for MobileOrg. Org also uses the same directory for sharing notes between Org and MobileOrg.

B.2 Pushing to MobileOrg

Org pushes files listed in `org-mobile-files` to `org-mobile-directory`. Files include agenda files (as listed in `org-agenda-files`). Customize `org-mobile-files` to add other files. File names will be staged with paths relative to `org-directory`, so all files should be inside this directory[3].

---

[1] If Emacs is configured for safe storing of passwords, then configure the variable, `org-mobile-encryption-password`; please read the docstring of that variable.

[2] An alternative is to use webdav server. MobileOrg documentation has details of webdav server configuration. Additional help is at FAQ entry.

[3] Symbolic links in `org-directory` should have the same name as their targets.

Push creates a special Org file **agendas.org** with custom agenda views defined by the user[4].

Org writes the file **index.org**, containing links to other files. MobileOrg reads this file first from the server to determine what other files to download for agendas. For faster downloads, MobileOrg will read only those files whose checksums[5] have changed.

B.3 Pulling from MobileOrg

When MobileOrg synchronizes with the server, it pulls the Org files for viewing. It then appends to the file `mobileorg.org` on the server the captured entries, pointers to flagged and changed entries. Org integrates its data in an inbox file format.

1. Org moves all entries found in `mobileorg.org`[6] and appends them to the file pointed to by the variable `org-mobile-inbox-for-pull`. Each captured entry and each editing event is a top-level entry in the inbox file.
2. After moving the entries, Org attempts changes to MobileOrg. Some changes are applied directly and without user interaction. Examples include changes to tags, TODO state, headline and body text. Entries for further action are tagged as `:FLAGGED:`. Org marks entries with problems with an error message in the inbox. They have to be resolved manually.
3. Org generates an agenda view for flagged entries for user intervention to clean up. For notes stored in flagged entries, MobileOrg displays them in the echo area when the cursor is on the corresponding agenda item.

   ? Pressing *?* displays the entire flagged note in another window. Org also pushes it to the kill ring. To store flagged note as a normal note, use *? z C-y C-c C-c*. Pressing *?* twice does these things: first it removes the `:FLAGGED:` tag; second, it removes the flagged note from the property drawer; third, it signals that manual editing of the flagged entry is now finished.

C-c a ? returns to the agenda view to finish processing flagged entries. Note that these entries may not be the most recent since MobileOrg searches files that were last pulled. To get an updated agenda view with changes since the last pull, pull again.

---

[4] While creating the agendas, Org mode will force ID properties on all referenced entries, so that these entries can be uniquely identified if MobileOrg flags them for further action. To avoid setting properties configure the variable `org-mobile-force-id-on-agenda-items` to `nil`. Org mode will then rely on outline paths, assuming they are unique.

[5] Checksums are stored automatically in the file `checksums.dat`.

[6] `mobileorg.org` will be empty after this operation.

Appendix C History and acknowledgments

C.1 From Carsten

Org was born in 2003, out of frustration over the user interface of the Emacs Outline mode. I was trying to organize my notes and projects, and using Emacs seemed to be the natural way to go. However, having to remember eleven different commands with two or three keys per command, only to hide and show parts of the outline tree, that seemed entirely unacceptable. Also, when using outlines to take notes, I constantly wanted to restructure the tree, organizing it paralleling my thoughts and plans. *Visibility cycling* and *structure editing* were originally implemented in the package `outline-magic.el`, but quickly moved to the more general `org.el`. As this environment became comfortable for project planning, the next step was adding *TODO entries*, basic *timestamps*, and *table support*. These areas highlighted the two main goals that Org still has today: to be a new, outline-based, plain text mode with innovative and intuitive editing features, and to incorporate project planning functionality directly into a notes file.

Since the first release, literally thousands of emails to me or to `emacs-orgmode@gnu.org` have provided a constant stream of bug reports, feedback, new ideas, and sometimes patches and add-on code. Many thanks to everyone who has helped to improve this package. I am trying to keep here a list of the people who had significant influence in shaping one or more aspects of Org. The list may not be complete, if I have forgotten someone, please accept my apologies and let me know.

Before I get to this list, a few special mentions are in order:

Bastien Guerry

Bastien has written a large number of extensions to Org (most of them integrated into the core by now), including the LaTeX exporter and the plain list parser. His support during the early days was central to the success of this project. Bastien also invented Worg, helped establishing the Web presence of Org, and sponsored hosting costs for the orgmode.org website. Bastien stepped in as maintainer of Org between 2011 and 2013, at a time when I desperately needed a break.

Eric Schulte and Dan Davison

Eric and Dan are jointly responsible for the Org-babel system, which turns Org into a multi-language environment for evaluating code and doing literate programming and reproducible research. This has become one of Org's killer features that define what Org is today.

John Wiegley

John has contributed a number of great ideas and patches directly to Org, including the attachment system (`org-attach.el`), integration with Apple Mail (`org-mac-message.el`), hierarchical dependencies of TODO items, habit tracking (`org-habits.el`), and encryption (`org-crypt.el`). Also, the capture system is really an extended copy of his great `remember.el`.

Sebastian Rose

Without Sebastian, the HTML/XHTML publishing of Org would be the pitiful work of an ignorant amateur. Sebastian has pushed this part of Org onto a much higher level. He also wrote `org-info.js`, a Java script for displaying web pages derived from Org using an Info-like or a folding interface with single-key navigation.

See below for the full list of contributions! Again, please let me know what I am missing here!

C.2 From Bastien

I (Bastien) have been maintaining Org between 2011 and 2013. This appendix would not be complete without adding a few more acknowledgments and thanks.

I am first grateful to Carsten for his trust while handing me over the maintainership of Org. His unremitting support is what really helped me getting more confident over time, with both the community and the code.

When I took over maintainership, I knew I would have to make Org more collaborative than ever, as I would have to rely on people that are more knowledgeable than I am on many parts of the code. Here is a list of the persons I could rely on, they should really be considered co-maintainers, either of the code or the community:

Eric Schulte
: Eric is maintaining the Babel parts of Org. His reactivity here kept me away from worrying about possible bugs here and let me focus on other parts.

Nicolas Goaziou
: Nicolas is maintaining the consistency of the deepest parts of Org. His work on `org-element.el` and `ox.el` has been outstanding, and it opened the doors for many new ideas and features. He rewrote many of the old exporters to use the new export engine, and helped with documenting this major change. More importantly (if that's possible), he has been more than reliable during all the work done for Org 8.0, and always very reactive on the mailing list.

Achim Gratz
: Achim rewrote the building process of Org, turning some *ad hoc* tools into a flexible and conceptually clean process. He patiently coped with the many hiccups that such a change can create for users.

Nick Dokos
: The Org mode mailing list would not be such a nice place without Nick, who patiently helped users so many times. It is impossible to overestimate such a great help, and the list would not be so active without him.

I received support from so many users that it is clearly impossible to be fair when shortlisting a few of them, but Org's history would not be complete if the ones above were not mentioned in this manual.

C.3 List of contributions

- *Russel Adams* came up with the idea for drawers.
- *Suvayu Ali* has steadily helped on the mailing list, providing useful feedback on many features and several patches.

- *Luis Anaya* wrote `ox-man.el`.
- *Thomas Baumann* wrote `org-bbdb.el` and `org-mhe.el`.
- *Michael Brand* helped by reporting many bugs and testing many features. He also implemented the distinction between empty fields and 0-value fields in Org's spreadsheets.
- *Christophe Bataillon* created the great unicorn logo that we use on the Org mode website.
- *Alex Bochannek* provided a patch for rounding timestamps.
- *Jan Böcker* wrote `org-docview.el`.
- *Brad Bozarth* showed how to pull RSS feed data into Org mode files.
- *Tom Breton* wrote `org-choose.el`.
- *Charles Cave*'s suggestion sparked the implementation of templates for Remember, which are now templates for capture.
- *Pavel Chalmoviansky* influenced the agenda treatment of items with specified time.
- *Gregory Chernov* patched support for Lisp forms into table calculations and improved XEmacs compatibility, in particular by porting `nouline.el` to XEmacs.
- *Sacha Chua* suggested copying some linking code from Planner, and helped make Org popular through her blog.
- *Toby S. Cubitt* contributed to the code for clock formats.
- *Baoqiu Cui* contributed the first DocBook exporter. In Org 8.0, we go a different route: you can now export to Texinfo and export the `.texi` file to DocBook using `makeinfo`.
- *Eddward DeVilla* proposed and tested checkbox statistics. He also came up with the idea of properties, and that there should be an API for them.
- *Nick Dokos* tracked down several nasty bugs.
- *Kees Dullemond* used to edit projects lists directly in HTML and so inspired some of the early development, including HTML export. He also asked for a way to narrow wide table columns.
- *Jason Dunsmore* has been maintaining the Org-Mode server at Rackspace for several years now. He also sponsored the hosting costs until Rackspace started to host us for free.
- *Thomas S. Dye* contributed documentation on Worg and helped integrating the Org-Babel documentation into the manual.
- *Christian Egli* converted the documentation into Texinfo format, inspired the agenda, patched CSS formatting into the HTML exporter, and wrote `org-taskjuggler.el`, which has been rewritten by Nicolas Goaziou as `ox-taskjuggler.el` for Org 8.0.
- *David Emery* provided a patch for custom CSS support in exported HTML agendas.
- *Sean Escriva* took over MobileOrg development on the iPhone platform.
- *Nic Ferrier* contributed mailcap and XOXO support.
- *Miguel A. Figueroa-Villanueva* implemented hierarchical checkboxes.
- *John Foerch* figured out how to make incremental search show context around a match in a hidden outline tree.

- *Raimar Finken* wrote `org-git-line.el`.
- *Mikael Fornius* works as a mailing list moderator.
- *Austin Frank* works as a mailing list moderator.
- *Eric Fraga* drove the development of BEAMER export with ideas and testing.
- *Barry Gidden* did proofreading the manual in preparation for the book publication through Network Theory Ltd.
- *Niels Giesen* had the idea to automatically archive DONE trees.
- *Nicolas Goaziou* rewrote much of the plain list code. He also wrote `org-element.el` and `org-export.el`, which was a huge step forward in implementing a clean framework for Org exporters.
- *Kai Grossjohann* pointed out key-binding conflicts with other packages.
- *Brian Gough* of Network Theory Ltd publishes the Org mode manual as a book.
- *Bernt Hansen* has driven much of the support for auto-repeating tasks, task state change logging, and the clocktable. His clear explanations have been critical when we started to adopt the Git version control system.
- *Manuel Hermenegildo* has contributed various ideas, small fixes and patches.
- *Phil Jackson* wrote `org-irc.el`.
- *Scott Jaderholm* proposed footnotes, control over whitespace between folded entries, and column view for properties.
- *Matt Jones* wrote *MobileOrg Android*.
- *Tokuya Kameshima* wrote `org-wl.el` and `org-mew.el`.
- *Jonathan Leech-Pepin* wrote `ox-texinfo.el`.
- *Shidai Liu* ("Leo") asked for embedded LaTeX and tested it. He also provided frequent feedback and some patches.
- *Matt Lundin* has proposed last-row references for table formulas and named invisible anchors. He has also worked a lot on the FAQ.
- *David Maus* wrote `org-atom.el`, maintains the issues file for Org, and is a prolific contributor on the mailing list with competent replies, small fixes and patches.
- *Jason F. McBrayer* suggested agenda export to CSV format.
- *Max Mikhanosha* came up with the idea of refiling and sticky agendas.
- *Dmitri Minaev* sent a patch to set priority limits on a per-file basis.
- *Stefan Monnier* provided a patch to keep the Emacs-Lisp compiler happy.
- *Richard Moreland* wrote MobileOrg for the iPhone.
- *Rick Moynihan* proposed allowing multiple TODO sequences in a file and being able to quickly restrict the agenda to a subtree.
- *Todd Neal* provided patches for links to Info files and Elisp forms.
- *Greg Newman* refreshed the unicorn logo into its current form.
- *Tim O'Callaghan* suggested in-file links, search options for general file links, and TAGS.
- *Osamu Okano* wrote `orgcard2ref.pl`, a Perl program to create a text version of the reference card.
- *Takeshi Okano* translated the manual and David O'Toole's tutorial into Japanese.

- *Oliver Oppitz* suggested multi-state TODO items.
- *Scott Otterson* sparked the introduction of descriptive text for links, among other things.
- *Pete Phillips* helped during the development of the TAGS feature, and provided frequent feedback.
- *Francesco Pizzolante* provided patches that helped speeding up the agenda generation.
- *Martin Pohlack* provided the code snippet to bundle character insertion into bundles of 20 for undo.
- *Rackspace.com* is hosting our website for free. Thank you Rackspace!
- *T.V. Raman* reported bugs and suggested improvements.
- *Matthias Rempe* (Oelde) provided ideas, Windows support, and quality control.
- *Paul Rivier* provided the basic implementation of named footnotes. He also acted as mailing list moderator for some time.
- *Kevin Rogers* contributed code to access VM files on remote hosts.
- *Frank Ruell* solved the mystery of the `keymapp nil` bug, a conflict with `allout.el`.
- *Jason Riedy* generalized the send-receive mechanism for Orgtbl tables with extensive patches.
- *Philip Rooke* created the Org reference card, provided lots of feedback, developed and applied standards to the Org documentation.
- *Christian Schlauer* proposed angular brackets around links, among other things.
- *Christopher Schmidt* reworked `orgstruct-mode` so that users can enjoy folding in non-org buffers by using Org headlines in comments.
- *Paul Sexton* wrote `org-ctags.el`.
- Linking to VM/BBDB/Gnus was first inspired by *Tom Shannon*'s `organizer-mode.el`.
- *Ilya Shlyakhter* proposed the Archive Sibling, line numbering in literal examples, and remote highlighting for referenced code lines.
- *Stathis Sideris* wrote the `ditaa.jar` ASCII to PNG converter that is now packaged into Org's `contrib` directory.
- *Daniel Sinder* came up with the idea of internal archiving by locking subtrees.
- *Dale Smith* proposed link abbreviations.
- *James TD Smith* has contributed a large number of patches for useful tweaks and features.
- *Adam Spiers* asked for global linking commands, inspired the link extension system, added support for mairix, and proposed the mapping API.
- *Ulf Stegemann* created the table to translate special symbols to HTML, LaTeX, UTF-8, Latin-1 and ASCII.
- *Andy Stewart* contributed code to `org-w3m.el`, to copy HTML content with links transformation to Org syntax.
- *David O'Toole* wrote `org-publish.el` and drafted the manual chapter about publishing.
- *Jambunathan K* contributed the ODT exporter and rewrote the HTML exporter.

- *Sebastien Vauban* reported many issues with LaTeX and BEAMER export and enabled source code highlighting in Gnus.
- *Stefan Vollmar* organized a video-recorded talk at the Max-Planck-Institute for Neurology. He also inspired the creation of a concept index for HTML export.
- *Jürgen Vollmer* contributed code generating the table of contents in HTML output.
- *Samuel Wales* has provided important feedback and bug reports.
- *Chris Wallace* provided a patch implementing the 'QUOTE' keyword.
- *David Wainberg* suggested archiving, and improvements to the linking system.
- *Carsten Wimmer* suggested some changes and helped fix a bug in linking to Gnus.
- *Roland Winkler* requested additional key bindings to make Org work on a tty.
- *Piotr Zielinski* wrote `org-mouse.el`, proposed agenda blocks and contributed various ideas and code snippets.
- *Marco Wahl* wrote `org-eww.el`.

Appendix D GNU Free Documentation License

Version 1.3, 3 November 2008

0. PREAMBLE

The purpose of this License is to make a manual, textbook, or other functional and useful document *free* in the sense of freedom: to assure everyone the effective freedom to copy and redistribute it, with or without modifying it, either commercially or noncommercially. Secondarily, this License preserves for the author and publisher a way to get credit for their work, while not being considered responsible for modifications made by others.

This License is a kind of "copyleft", which means that derivative works of the document must themselves be free in the same sense. It complements the GNU General Public License, which is a copyleft license designed for free software.

We have designed this License in order to use it for manuals for free software, because free software needs free documentation: a free program should come with manuals providing the same freedoms that the software does. But this License is not limited to software manuals; it can be used for any textual work, regardless of subject matter or whether it is published as a printed book. We recommend this License principally for works whose purpose is instruction or reference.

1. APPLICABILITY AND DEFINITIONS

This License applies to any manual or other work, in any medium, that contains a notice placed by the copyright holder saying it can be distributed under the terms of this License. Such a notice grants a world-wide, royalty-free license, unlimited in duration, to use that work under the conditions stated herein. The "Document", below, refers to any such manual or work. Any member of the public is a licensee, and is addressed as "you". You accept the license if you copy, modify or distribute the work in a way requiring permission under copyright law.

A "Modified Version" of the Document means any work containing the Document or a portion of it, either copied verbatim, or with modifications and/or translated into another language.

A "Secondary Section" is a named appendix or a front-matter section of the Document that deals exclusively with the relationship of the publishers or authors of the Document to the Document's overall subject (or to related matters) and contains nothing that could fall directly within that overall subject. (Thus, if the Document is in part a textbook of mathematics, a Secondary Section may not explain any mathematics.) The relationship could be a matter of historical connection with the subject or with related matters, or of legal, commercial, philosophical, ethical or political position regarding them.

The "Invariant Sections" are certain Secondary Sections whose titles are designated, as being those of Invariant Sections, in the notice that says that the Document is released under this License. If a section does not fit the above definition of Secondary then it is not allowed to be designated as Invariant. The Document may contain zero Invariant Sections. If the Document does not identify any Invariant Sections then there are none.

The "Cover Texts" are certain short passages of text that are listed, as Front-Cover Texts or Back-Cover Texts, in the notice that says that the Document is released under this License. A Front-Cover Text may be at most 5 words, and a Back-Cover Text may be at most 25 words.

A "Transparent" copy of the Document means a machine-readable copy, represented in a format whose specification is available to the general public, that is suitable for revising the document straightforwardly with generic text editors or (for images composed of pixels) generic paint programs or (for drawings) some widely available drawing editor, and that is suitable for input to text formatters or for automatic translation to a variety of formats suitable for input to text formatters. A copy made in an otherwise Transparent file format whose markup, or absence of markup, has been arranged to thwart or discourage subsequent modification by readers is not Transparent. An image format is not Transparent if used for any substantial amount of text. A copy that is not "Transparent" is called "Opaque".

Examples of suitable formats for Transparent copies include plain ASCII without markup, Texinfo input format, LaTeX input format, SGML or XML using a publicly available DTD, and standard-conforming simple HTML, PostScript or PDF designed for human modification. Examples of transparent image formats include PNG, XCF and JPG. Opaque formats include proprietary formats that can be read and edited only by proprietary word processors, SGML or XML for which the DTD and/or processing tools are not generally available, and the machine-generated HTML, PostScript or PDF produced by some word processors for output purposes only.

The "Title Page" means, for a printed book, the title page itself, plus such following pages as are needed to hold, legibly, the material this License requires to appear in the title page. For works in formats which do not have any title page as such, "Title Page" means the text near the most prominent appearance of the work's title, preceding the beginning of the body of the text.

The "publisher" means any person or entity that distributes copies of the Document to the public.

A section "Entitled XYZ" means a named subunit of the Document whose title either is precisely XYZ or contains XYZ in parentheses following text that translates XYZ in another language. (Here XYZ stands for a specific section name mentioned below, such as "Acknowledgements", "Dedications", "Endorsements", or "History".) To "Preserve the Title" of such a section when you modify the Document means that it remains a section "Entitled XYZ" according to this definition.

The Document may include Warranty Disclaimers next to the notice which states that this License applies to the Document. These Warranty Disclaimers are considered to be included by reference in this License, but only as regards disclaiming warranties: any other implication that these Warranty Disclaimers may have is void and has no effect on the meaning of this License.

2. VERBATIM COPYING

You may copy and distribute the Document in any medium, either commercially or noncommercially, provided that this License, the copyright notices, and the license notice saying this License applies to the Document are reproduced in all copies, and that you add no other conditions whatsoever to those of this License. You may not use technical measures to obstruct or control the reading or further copying of the copies you make or distribute. However, you may accept compensation in exchange for copies. If you distribute a large enough number of copies you must also follow the conditions in section 3.

You may also lend copies, under the same conditions stated above, and you may publicly display copies.

3. COPYING IN QUANTITY

If you publish printed copies (or copies in media that commonly have printed covers) of the Document, numbering more than 100, and the Document's license notice requires Cover Texts, you must enclose the copies in covers that carry, clearly and legibly, all these Cover Texts: Front-Cover Texts on the front cover, and Back-Cover Texts on the back cover. Both covers must also clearly and legibly identify you as the publisher of these copies. The front cover must present the full title with all words of the title equally prominent and visible. You may add other material on the covers in addition. Copying with changes limited to the covers, as long as they preserve the title of the Document and satisfy these conditions, can be treated as verbatim copying in other respects.

If the required texts for either cover are too voluminous to fit legibly, you should put the first ones listed (as many as fit reasonably) on the actual cover, and continue the rest onto adjacent pages.

If you publish or distribute Opaque copies of the Document numbering more than 100, you must either include a machine-readable Transparent copy along with each Opaque copy, or state in or with each Opaque copy a computer-network location from which the general network-using public has access to download using public-standard network protocols a complete Transparent copy of the Document, free of added material. If you use the latter option, you must take reasonably prudent steps, when you begin distribution of Opaque copies in quantity, to ensure that this Transparent copy will remain thus accessible at the stated location until at least one year after the last time you distribute an Opaque copy (directly or through your agents or retailers) of that edition to the public.

It is requested, but not required, that you contact the authors of the Document well before redistributing any large number of copies, to give them a chance to provide you with an updated version of the Document.

4. MODIFICATIONS

You may copy and distribute a Modified Version of the Document under the conditions of sections 2 and 3 above, provided that you release the Modified Version under precisely this License, with the Modified Version filling the role of the Document, thus licensing distribution and modification of the Modified Version to whoever possesses a copy of it. In addition, you must do these things in the Modified Version:

A. Use in the Title Page (and on the covers, if any) a title distinct from that of the Document, and from those of previous versions (which should, if there were any, be listed in the History section of the Document). You may use the same title as a previous version if the original publisher of that version gives permission.

B. List on the Title Page, as authors, one or more persons or entities responsible for authorship of the modifications in the Modified Version, together with at least five of the principal authors of the Document (all of its principal authors, if it has fewer than five), unless they release you from this requirement.

C. State on the Title page the name of the publisher of the Modified Version, as the publisher.

D. Preserve all the copyright notices of the Document.

E. Add an appropriate copyright notice for your modifications adjacent to the other copyright notices.

F. Include, immediately after the copyright notices, a license notice giving the public permission to use the Modified Version under the terms of this License, in the form shown in the Addendum below.

G. Preserve in that license notice the full lists of Invariant Sections and required Cover Texts given in the Document's license notice.

H. Include an unaltered copy of this License.

I. Preserve the section Entitled "History", Preserve its Title, and add to it an item stating at least the title, year, new authors, and publisher of the Modified Version as given on the Title Page. If there is no section Entitled "History" in the Document, create one stating the title, year, authors, and publisher of the Document as given on its Title Page, then add an item describing the Modified Version as stated in the previous sentence.

J. Preserve the network location, if any, given in the Document for public access to a Transparent copy of the Document, and likewise the network locations given in the Document for previous versions it was based on. These may be placed in the "History" section. You may omit a network location for a work that was published at least four years before the Document itself, or if the original publisher of the version it refers to gives permission.

K. For any section Entitled "Acknowledgements" or "Dedications", Preserve the Title of the section, and preserve in the section all the substance and tone of each of the contributor acknowledgements and/or dedications given therein.

L. Preserve all the Invariant Sections of the Document, unaltered in their text and in their titles. Section numbers or the equivalent are not considered part of the section titles.

M. Delete any section Entitled "Endorsements". Such a section may not be included in the Modified Version.

N. Do not retitle any existing section to be Entitled "Endorsements" or to conflict in title with any Invariant Section.

O. Preserve any Warranty Disclaimers.

If the Modified Version includes new front-matter sections or appendices that qualify as Secondary Sections and contain no material copied from the Document, you may at your option designate some or all of these sections as invariant. To do this, add their titles to the list of Invariant Sections in the Modified Version's license notice. These titles must be distinct from any other section titles.

You may add a section Entitled "Endorsements", provided it contains nothing but endorsements of your Modified Version by various parties—for example, statements of peer review or that the text has been approved by an organization as the authoritative definition of a standard.

You may add a passage of up to five words as a Front-Cover Text, and a passage of up to 25 words as a Back-Cover Text, to the end of the list of Cover Texts in the Modified Version. Only one passage of Front-Cover Text and one of Back-Cover Text may be added by (or through arrangements made by) any one entity. If the Document already includes a cover text for the same cover, previously added by you or by arrangement made by the same entity you are acting on behalf of, you may not add another; but you may replace the old one, on explicit permission from the previous publisher that added the old one.

The author(s) and publisher(s) of the Document do not by this License give permission to use their names for publicity for or to assert or imply endorsement of any Modified Version.

5. COMBINING DOCUMENTS

You may combine the Document with other documents released under this License, under the terms defined in section 4 above for modified versions, provided that you include in the combination all of the Invariant Sections of all of the original documents, unmodified, and list them all as Invariant Sections of your combined work in its license notice, and that you preserve all their Warranty Disclaimers.

The combined work need only contain one copy of this License, and multiple identical Invariant Sections may be replaced with a single copy. If there are multiple Invariant Sections with the same name but different contents, make the title of each such section unique by adding at the end of it, in parentheses, the name of the original author or publisher of that section if known, or else a unique number. Make the same adjustment to the section titles in the list of Invariant Sections in the license notice of the combined work.

In the combination, you must combine any sections Entitled "History" in the various original documents, forming one section Entitled "History"; likewise combine any sections Entitled "Acknowledgements", and any sections Entitled "Dedications". You must delete all sections Entitled "Endorsements."

6. COLLECTIONS OF DOCUMENTS

You may make a collection consisting of the Document and other documents released under this License, and replace the individual copies of this License in the various documents with a single copy that is included in the collection, provided that you follow the rules of this License for verbatim copying of each of the documents in all other respects.

You may extract a single document from such a collection, and distribute it individually under this License, provided you insert a copy of this License into the extracted document, and follow this License in all other respects regarding verbatim copying of that document.

7. AGGREGATION WITH INDEPENDENT WORKS

   A compilation of the Document or its derivatives with other separate and independent documents or works, in or on a volume of a storage or distribution medium, is called an "aggregate" if the copyright resulting from the compilation is not used to limit the legal rights of the compilation's users beyond what the individual works permit. When the Document is included in an aggregate, this License does not apply to the other works in the aggregate which are not themselves derivative works of the Document.

   If the Cover Text requirement of section 3 is applicable to these copies of the Document, then if the Document is less than one half of the entire aggregate, the Document's Cover Texts may be placed on covers that bracket the Document within the aggregate, or the electronic equivalent of covers if the Document is in electronic form. Otherwise they must appear on printed covers that bracket the whole aggregate.

8. TRANSLATION

   Translation is considered a kind of modification, so you may distribute translations of the Document under the terms of section 4. Replacing Invariant Sections with translations requires special permission from their copyright holders, but you may include translations of some or all Invariant Sections in addition to the original versions of these Invariant Sections. You may include a translation of this License, and all the license notices in the Document, and any Warranty Disclaimers, provided that you also include the original English version of this License and the original versions of those notices and disclaimers. In case of a disagreement between the translation and the original version of this License or a notice or disclaimer, the original version will prevail.

   If a section in the Document is Entitled "Acknowledgements", "Dedications", or "History", the requirement (section 4) to Preserve its Title (section 1) will typically require changing the actual title.

9. TERMINATION

   You may not copy, modify, sublicense, or distribute the Document except as expressly provided under this License. Any attempt otherwise to copy, modify, sublicense, or distribute it is void, and will automatically terminate your rights under this License.

   However, if you cease all violation of this License, then your license from a particular copyright holder is reinstated (a) provisionally, unless and until the copyright holder explicitly and finally terminates your license, and (b) permanently, if the copyright holder fails to notify you of the violation by some reasonable means prior to 60 days after the cessation.

   Moreover, your license from a particular copyright holder is reinstated permanently if the copyright holder notifies you of the violation by some reasonable means, this is the first time you have received notice of violation of this License (for any work) from that copyright holder, and you cure the violation prior to 30 days after your receipt of the notice.

   Termination of your rights under this section does not terminate the licenses of parties who have received copies or rights from you under this License. If your rights have been terminated and not permanently reinstated, receipt of a copy of some or all of the same material does not give you any rights to use it.

10. FUTURE REVISIONS OF THIS LICENSE

    The Free Software Foundation may publish new, revised versions of the GNU Free Documentation License from time to time. Such new versions will be similar in spirit to the present version, but may differ in detail to address new problems or concerns. See `http://www.gnu.org/copyleft/`.

    Each version of the License is given a distinguishing version number. If the Document specifies that a particular numbered version of this License "or any later version" applies to it, you have the option of following the terms and conditions either of that specified version or of any later version that has been published (not as a draft) by the Free Software Foundation. If the Document does not specify a version number of this License, you may choose any version ever published (not as a draft) by the Free Software Foundation. If the Document specifies that a proxy can decide which future versions of this License can be used, that proxy's public statement of acceptance of a version permanently authorizes you to choose that version for the Document.

11. RELICENSING

    "Massive Multiauthor Collaboration Site" (or "MMC Site") means any World Wide Web server that publishes copyrightable works and also provides prominent facilities for anybody to edit those works. A public wiki that anybody can edit is an example of such a server. A "Massive Multiauthor Collaboration" (or "MMC") contained in the site means any set of copyrightable works thus published on the MMC site.

    "CC-BY-SA" means the Creative Commons Attribution-Share Alike 3.0 license published by Creative Commons Corporation, a not-for-profit corporation with a principal place of business in San Francisco, California, as well as future copyleft versions of that license published by that same organization.

    "Incorporate" means to publish or republish a Document, in whole or in part, as part of another Document.

    An MMC is "eligible for relicensing" if it is licensed under this License, and if all works that were first published under this License somewhere other than this MMC, and subsequently incorporated in whole or in part into the MMC, (1) had no cover texts or invariant sections, and (2) were thus incorporated prior to November 1, 2008.

    The operator of an MMC Site may republish an MMC contained in the site under CC-BY-SA on the same site at any time before August 1, 2009, provided the MMC is eligible for relicensing.

ADDENDUM: How to use this License for your documents

To use this License in a document you have written, include a copy of the License in the document and put the following copyright and license notices just after the title page:

```
Copyright (C)  year  your name.
Permission is granted to copy, distribute and/or modify this document
under the terms of the GNU Free Documentation License, Version 1.3
or any later version published by the Free Software Foundation;
with no Invariant Sections, no Front-Cover Texts, and no Back-Cover
Texts.  A copy of the license is included in the section entitled ``GNU
Free Documentation License''.
```

If you have Invariant Sections, Front-Cover Texts and Back-Cover Texts, replace the "with...Texts." line with this:

```
with the Invariant Sections being list their titles, with
the Front-Cover Texts being list, and with the Back-Cover Texts
being list.
```

If you have Invariant Sections without Cover Texts, or some other combination of the three, merge those two alternatives to suit the situation.

If your document contains nontrivial examples of program code, we recommend releasing these examples in parallel under your choice of free software license, such as the GNU General Public License, to permit their use in free software.

Concept index

#

%

:

A

B

C

D

E

F

G

M

N

O

P

Q

R

S

T

U

V

W

Y

Z

Key index

S

T

U

V

W

X

Z

Command and function index

L

N

O

P

W

Variable index

This is not a complete index of variables and faces, only the ones that are mentioned in the manual. For a complete list, use *M-x org-customize RET*.

C

H

L

O

P

U

CPSIA information can be obtained
at www.ICGtesting.com
Printed in the USA
LVHW101139030419
612814LV00018B/226/P